Sebastian Bauer

Hyperspectral Image Unmixing Incorporating Adjacency Information

Forschungsberichte aus der Industriellen Informationstechnik
Band 18

Institut für Industrielle Informationstechnik
Karlsruher Institut für Technologie
Hrsg. Prof. Dr.-Ing. Fernando Puente León
Prof. Dr.-Ing. habil. Klaus Dostert

Eine Übersicht aller bisher in dieser Schriftenreihe erschienenen Bände finden Sie am Ende des Buchs.

Hyperspectral Image Unmixing Incorporating Adjacency Information

by
Sebastian Bauer

KIT Scientific Publishing

Karlsruher Institut für Technologie
Institut für Industrielle Informationstechnik

Hyperspectral Image Unmixing Incorporating Adjacency Information

Zur Erlangung des akademischen Grades eines Doktor-Ingenieurs von der KIT-Fakultät für Elektrotechnik und Informationstechnik des Karlsruher Instituts für Technologie (KIT) genehmigte Dissertation

von Sebastian Bernd Bauer, M.Sc., geb. in Nürnberg

Tag der mündlichen Prüfung: 16. April 2018
Hauptreferent: Prof. Dr.-Ing. Fernando Puente León, KIT
Korreferent: Prof. Dr. rer. nat. Bernd Jähne, Universität Heidelberg

Impressum

KIT Scientific Publishing

Karlsruher Institut für Technologie (KIT)
KIT Scientific Publishing
Straße am Forum 2
D-76131 Karlsruhe

KIT Scientific Publishing is a registered trademark
of Karlsruhe Institute of Technology.
Reprint using the book cover is not allowed.

www.ksp.kit.edu

Print on Demand 2018 – Gedruckt auf FSC-zertifiziertem Papier

ISSN 2190-6629
ISBN 978-3-7315-0788-8
DOI 10.5445/KSP/1000081665

Hyperspectral Image Unmixing Incorporating Adjacency Information

Zur Erlangung des akademischen Grades eines

DOKTOR-INGENIEURS

von der Fakultät für

Elektrotechnik und Informationstechnik

des Karlsruher Instituts für Technologie (KIT)

genehmigte

DISSERTATION

von

Sebastian Bernd Bauer, M.Sc.

geb. in Nürnberg

Tag der mündl. Prüfung: 16. April 2018
Hauptreferent: Prof. Dr.-Ing. Fernando Puente León, KIT
Korreferent: Prof. Dr. rer. nat. Bernd Jähne, Universität Heidelberg

Preface and Acknowledgment

While it is the researchers who do the work, there is lots of financial and organizational effort necessary such that they are able to work properly. For this reason, I want to thank my supervisor, Prof. Dr.-Ing. Fernando Puente León, for giving me the opportunity to work at his institute, for his extraordinary support and for the opportunities he gave me that still continue to let me grow.

Within the environment he provided, many people made the achievements of this thesis possible. Miss Manuela Moritz, Mr. Dieter Brandt and Mr. Marvin Winkler gave me administrative support, and Mr. Stefan Ziegler and his team of the mechanical workshop produced the required high quality parts and setups.

It is a great honor for me that Prof. Dr. rer. nat. Bernd Jähne agreed to be the second examiner of my thesis. I also want to thank Prof. Dr. rer. nat. Olaf Dössel, Prof. Dr.-Ing. habil. Gert F. Trommer and Prof. Dr.-Ing. Eric Sax for their service on the board of examiners.

While my work on hyperspectral imaging has been a unique field of resarch at the IIIT for quite some time, I am extremely grateful to the department Visual Inspection Systems of the Fraunhofer Institute of Optronics, System Technologies and Image Exploitation for our collaboration. Special thanks go to the head, Prof. Dr.-Ing. Thomas Längle, as well as Dr. Miro Taphanel, Christian Negara, Dr. Kai-Uwe Vieth, Henning Schulte, Dr. Robin Gruna, Julius Krause, Johannes Meyer, Matthias Richter and Jürgen Hock for numerous discussions and help. Additionally, I was provided with easy access to lab equipment and also to data. Results obtained from these are shown in this thesis. Furthermore, I want to thank Britta Ost and Petra Riegel for their support, and the whole department for the invitation to many of their events.

I have had the honor to supervise many student theses while working on my PhD. Naming the most fruitful theses that also contributed to the present document, I want to thank Patrick Mackens, Florian Neumann,

Minh Hai Phan, Johannes Stefan, Felix Wagner and Eric Winterbauer for their great work.

Furthermore, I want to thank my fellow PhD students for the nice working atmosphere; special thanks go to Thomas Nürnberg, Sebastian Vater, Johannes Pallauf, Oliver Opalko, Cornelius Kaiser, Damián Alonso and Wolfgang Krippner. We also had fun while not at work.

I gratefully acknowledge support by the state of Baden-Württemberg through bwHPC. Many calculations were run on the BwUniCluster. Furthermore, I want to thank my desktop computer IIIT-Fourier for his continuous support without breaking down, especially in times of great computational load.

My deepest respect, gratitude and love is due to my parents, for their loving upbringing and their mental, physical and financial support not only during the work on this thesis, but my whole life. You are definitely outstanding idols. Also, I want to express my gratitude to my brother Michael who gained lots of experience while going a very similar way – thank you for clearing a lot of obstacles and providing an easier path!

Karlsruhe, April 2018 Sebastian Bernd Bauer

To the universe – and eternal peace on earth

Contents

Nomenclature

Common abbreviations

Abbreviation	Description
3DDCT-NIMF	3D discrete cosine transform nonnegative inverse matrix factorization
3DSVD-NIMF	3D SVD nonnegative inverse matrix factorization
3D DCT	Three-dimensional discrete cosine transform
ABU RMSE	Abundance RMSE
AWGN	Additive white Gaussian noise
BSS	Blind source separation
CIC	Clock induced charge
DCT	Discrete cosine transform
DFT	Discrete Fourier transform
ED	Euclidean distance
EEA	Endmember extraction algorithm
EM RMSE	Endmember RMSE
EMCCD	Electron-multiplying charge-coupled device
EM	Endmember
FS	UNCLI final step
FWHM	Full width half maximum
HSEG	Hierarchical segmentation
HS	Spatial second derivative regularization: Hessian regularization
HSI	Hyperspectral imaging
HSSNR	Hybrid spatial-spectral noise reduction
HYDICE	Hyperspectral digital imagery collection experiment
ICA	Independent component analysis

Abbreviation	Description
ISODATA	Iterative self-organizing data analysis technique
LA	Local averaging
LMM	Linear mixing model
MDC-NMF	Minimum distance constrained NMF
MewDC-NMF	Minimum endmember-wise distance constrained NMF
MIDWAI	Multidimensional image denoising by weighted addition of images
MiniDisCo-NMF	Minimum dispersion constrained NMF
MNF	Maximum noise fraction
MPSNR	Mean peak signal-to-noise ratio
MSE	Mean squared error
MVC-NMF	Minimum volume constrained NMF
MVES	Minimum volume enclosing simplex
MVSA	Minimum volume simplex analysis
MVT	Minimum volume transform
NAILRMA	Noise adjusted iterative low-rank matrix approximation
NIMF	Nonnegative inverse matrix factorization
NIR	Near-infrared
NMF	Nonnegative matrix factorization
NN constraint	Nonnegativity constraint
OSP	Orthogonal subspace projection
PCA	Principal component analysis
PDF	Probability density function
PQN	Projected quasi-Newton
PSNR	Peak signal-to-noise ratio
PYRU	Pyramid unmixing
RGB	Red Green Blue
RHSEG	Recursive hierarchical segmentation
RMSE	Root-mean-square error
SA	Spectral angle
SISAL	Simplex identification via split augmented Lagrangian
SNR	Signal-to-noise ratio

Abbreviation	Description
SSAHTV	Spectral–spatial adaptive hyperspectral TV model
STO constraint	Sum-to-one constraint
SU	Spectral unmixing
SVD	Singular value decomposition
SWIR	Short-wavelength infrared
TV	Total variation
UNCLI	Unmixing of clustered image
VCA	Vertex component analysis

Letters

Latin Letters

Symbol	Description
$\mathbf{0}$	Matrix of zeros
$\mathbb{1}$	Matrix of ones
$\mathbf{A}$	Abundance matrix, $\mathbf{A} \in \mathbb{R}^{R \times I}$
$\mathbf{A}_r$	Abundance map of r-th endmember, i.e., r-th row of $\mathbf{A}$ rearranged to original image size: $\mathbf{A}_r \in \mathbb{R}^{X \times Y}$
$\mathbf{A}_r(x,y)$	Abundance map $\mathbf{A}_r$ at position (x,y)
$\mathbf{A}_{r,xx}(x,y)$	Second derivative of the abundance map $\mathbf{A}_r$ in the direction of x at position (x,y)
$\mathbf{A}^k$	Abundance matrix on PYRU level k; $k=0$ denotes full resolution
$\mathcal{A}$	Abundance tensor, $\mathcal{A} \in \mathbb{R}^{X \times Y \times R}$
$\mathcal{A}^k$	Abundance tensor on PYRU level k; $k=0$ denotes full resolution
a	Analog-to-digital proportionality factor of an EMCCD
$\mathbf{a}_i$	Abundance vector of pixel i
B	PYRU iteration formula base, and 3DDCT-NIMF parameter defining the number of frequency coefficients considered in the final step
$\mathbf{B}$	Matrix of constants for MIDWAI denoising
C	Number of pixel clusters
$\mathbf{C}$	Matrix of constants for MIDWAI denoising
$\mathbf{C}_\mathbf{Y}$	Clustering segmentation map, $\mathbf{C}_\mathbf{Y} \in \mathbb{R}^{X \times Y}$
$\mathbf{C}^k$	PYRU segmentation map
c	Mean of clock induced charge of EMCCD sensor
$c_{\mathbf{x},\mathbf{f}}$	Constant encoding the 3D DCT cosine terms
d	Image sensor dark current rate in electrons per pixel per second
E	3DDCT-NIMF regularization parameter exponent
$F_{\mathtt{W}}(W)$	Characteristic function of $\mathtt{W}$
f_l	Index of spectral frequency axis

Symbol	Description
f_x	Index of first spatial frequency axis
f_y	Index of second spatial frequency axis
$f_{\mathtt{w}}(w)$	PDF of random variable $\mathtt{w}$
$\mathbf{f}$	DCT domain coordinate vector; $\mathbf{f} = [f_x, f_y, f_l]^{\mathrm{T}}$
G	EMCCD gain
g	Image value at one pixel
$\mathbf{g}$	3DDCT-NIMF scaling vector
I	Number of spatial pixels, $I = x \cdot y$
i	Pixel index
$\mathtt{i}$	Number of incident photons at the sensor
J	Number of matrix blocks in 3DDCT-NIMF
$J_{\mathrm{relchange}}$	Optimization stopping parameter
j	3DDCT-NIMF matrix block index
K	PYRU index of lowest resolution stage
K_j	3DDCT-NIMF block dependent regularization parameter
k	PYRU pyramid stage index; in different contexts: arbitrary constant value
L	Number of spectral bands
L'	Number of spectral bands after dimensionality reduction
l	Index of spectral image dimension
M	Number of endmember iterations
$\mathbf{M}$	Endmember matrix, $\mathbf{M} \in \mathbb{R}^{L \times R}$
$\mathbf{M}_r$	Spectrum of the r-th endmember; r-th column of $\mathbf{M}$, $\mathbf{M}_r \in \mathbb{R}^{1 \times L}$
m	Endmember iteration index and Armijo rule step size decrease iteration number
N	PYRU spatial reduction factor
$\mathbf{N}_j$	Matrix of j-th inverse endmember block in DCT domain; $\mathbf{N}_j \in \mathbb{R}^{R \times R}$
$\mathtt{n}$	Photon count; number of photons generating photoelectrons at one pixel during the exposure time

Symbol	Description
P	Number of iterations, typically abundance iterations. In context of 3DDCT-NIMF: constant in regularization parameter.
$P_{\mathrm{p}}(\lambda)$	Poisson distribution with parameter λ.
$\mathbf{P}$	Matrix containing the $R-1$ most significant spectral principal components of $\mathbf{Y}$, $\mathbf{P} \in \mathbb{R}^{L\times(R-1)}$
p	Current iteration, typically when abundance iterations are considered
Q	Objective function
Q_{data}	Data fidelity
R	Number of endmembers
r	Endmember index
$\mathcal{R}_{\mathrm{spat}}$	Spatial regularizer
$\mathcal{R}_{\mathrm{spec}}$	Spectral regularizer
$\mathbf{S}$	Subgradient matrix
s	Armijo rule initial step size
$\mathbf{s}_i$	Subgradient vector of pixel i
sw	Scalar spectral weight parameter
T	Total number of abundance/endmember sequences
t	Current iteration
$\mathbf{U}$	Denoised image matrix of 2D image or orthogonal SVD matrix
V	PYRU iteration formula prefactor
$\mathbf{V}$	3DDCT filter matrix or orthogonal SVD matrix
$\mathbf{W}_X$	1D DCT transform matrix of size $X \times X$
$\mathbf{W}_{XY}$	2D DCT transform matrix for unfolded image
$\mathtt{w}$	Random variable $\mathtt{w}$
X	Number of pixels along first spatial image dimension
x	Index of first spatial image dimension
$\mathbf{x}$	Hyperspectral image coordinate vector, $\mathbf{x} = [x, y, l]^{\mathrm{T}}$
Y	Number of pixels along second spatial image dimension
$\mathbf{Y}$	Unfolded image matrix, $\mathbf{Y} \in \mathbb{R}^{L\times I}$, where $I = X \cdot Y$
$\mathbf{Y}_{\mathrm{3DDCT}}$	Unfolded 3D DCT tensor, $\mathbf{Y}_{\mathrm{3DDCT}} \in \mathbb{R}^{L\times I}$

Symbol	Description
$\mathbf{Y}_{3\mathrm{DDCT},j}$	j-th block of the unfolded 3D DCT tensor, $\mathbf{Y}_{3\mathrm{DDCT},j} \in \mathbb{R}^{R\times I}$
$\mathtt{Y}_l$	l-th band tensor of Poisson distributed random variables, $\mathtt{Y}_l \in \mathbb{R}^{X\times Y}$
$\mathbf{Y}_l$	l-th band of $\mathcal{Y}$, $\mathbf{Y}_l \in \mathbb{R}^{X\times Y}$
$\mathbf{Y}^k$	Image matrix on PYRU level k; $k=0$ denotes full resolution
$\mathbf{Y}'$	Unfolded image matrix after spectral dimensionality reduction, $\mathbf{Y} \in \mathbb{R}^{L'\times I}$
$\mathcal{Y}$	Hyperspectral image tensor, $\mathcal{Y} \in \mathbb{R}^{X\times Y\times L}$
$\mathcal{Y}_{3\mathrm{DDCT}}$	3D DCT transformed hyperspectral image tensor, $\mathcal{Y}_{3\mathrm{DDCT}} \in \mathbb{R}^{X\times Y\times L}$
$\mathcal{Y}_{3\mathrm{DDCT},j}$	j-th block of $\mathcal{Y}_{3\mathrm{DDCT}}$, $\mathcal{Y}_{3\mathrm{DDCT},j} \in \mathbb{R}^{X\times Y\times R}$
$\mathcal{Y}^k$	Image tensor on PYRU level k; $k=0$ denotes full resolution
y	Index of second spatial image dimension
$y_{3\mathrm{DDCT},\mathbf{f}}$	Frequency coefficient at position $\mathbf{f}$ of $\mathbf{Y}_{3\mathrm{DDCT}}$
$y_{\mathbf{x}}$	Image element at position $\mathbf{x}$ of $\mathcal{Y}$
$\mathbf{y}_i$	i-th column of $\mathbf{Y}$; spectrum of image pixel i
Z	MIDWAI: number of neighboring bands taken into account
$\mathbf{Z}$	Measured image matrix of 2D image
$\mathtt{z}_{\mathbf{f}}$	Noise random variable of frequency coefficient $y_{3\mathrm{DDCT},\mathbf{f}}$
$\mathtt{z}_{\mathbf{x}}$	Noise random variable of image value $y_{\mathbf{x}}$

Greek Letters

Symbol	Description
α	Image sensor quantum efficiency
β	Armijo rule step size decrease parameter
γ	Algorithm parameter
Δt	Exposure time
δ	Gradient descent or subgradient step size

Symbol	Description
$\boldsymbol{\Lambda}_l$	Mean matrix of 2D matrix of random variables of band l
λ	Mean and variance of Poisson distributed random variable/matrix eigenvalue (meaning is clear from respective context)
λ_{em}	Fluorescence emission wavelength
λ_{exc}	Fluorescence excitation wavelength
λ_{EMCCD}	Number of electrons in EMCCD
μ	Arbitrary regularization parameter
μ_{spat}	Spatial regularization parameter
μ_{spec}	Spectral regularization parameter
$\boldsymbol{\mu}$	Spectral mean vector of $\mathbf{Y}$
ρ	Ratio of the largest absolute frequency coefficients that is retained
$\boldsymbol{\Sigma}$	Diagonal matrix of singular values
σ	Standard deviation of normal distribution/Armijo rule sufficient decrease parameter/matrix singular value (meaning clear from context)
σ_{GK}	Standard deviation of MIDWAI Gaussian kernel
$\boldsymbol{\psi}$	Noise and model error vector
$\boldsymbol{\Psi}$	Noise and model error matrix

Mathematical operators

Operator	Description
$\lceil\cdot\rceil$	Ceiling function; smallest integer larger than or equal to the argument
$\lfloor\cdot\rfloor$	Floor function; largest integer smaller than or equal to the argument
$\otimes$	Kronecker product
$\odot$	Hadamard matrix product (elementwise multiplication)
$(\cdot).^2$	Matrix elementwise squaring
$\lVert\cdot\rVert_2$	Spectral norm (maximum singular value)

Operator	Description
$\lVert\cdot\rVert_{\mathrm{F}}$	Frobenius norm
$(\cdot)_{\mathrm{P}}$	Projection onto a convex set
$\lVert\cdot\rVert_p$	Vector ℓ_p norm or matrix p-norm
$(\cdot)_+$	Setting negative matrix elements to zero
$\mathrm{atr}\,(\cdot)$	Matrix absolute trace (sum of absolute diagonal elements)
$\det\,(\cdot)$	Matrix determinant
$\mathrm{diag}(\mathbf{x})$	Diagonal matrix containing the elements of vector $\mathbf{x}$ on its diagonal
$\mathrm{E}\{\cdot\}$	Expected value
$H(\cdot)(x, y)$	Hessian of a spatial matrix $(\cdot)$ at position (x, y)
$\log\,(\cdot)$	Natural logarithm
$\mathcal{N}(i)$	Neighborhood around pixel i
$\mathrm{nnz}(\cdot)$	Number of nonzero matrix elements
$\mathrm{sgn}\,(\cdot)$	Sign function
$\mathrm{tr}\,(\cdot)$	Matrix trace (sum of diagonal elements)
$\mathrm{Var}\{\cdot\}$	Variance
$\mathrm{vec}(\cdot)$	Matrix vectorization; stacking all columns on top of each other

1 Introduction

1.1 Hyperspectral Imaging

Common color cameras acquire an image of a scene by integrating the visible electromagnetic spectrum within three overlapping wavelength regions. Extensions of this concept, so-called multispectral cameras, have more than three color channels. Each of these channels still covers a more or less wide wavelength range. Hyperspectral cameras, by contrast, sample the spectrum at up to several hundred evenly spaced wavelengths (Fig. 1.1).

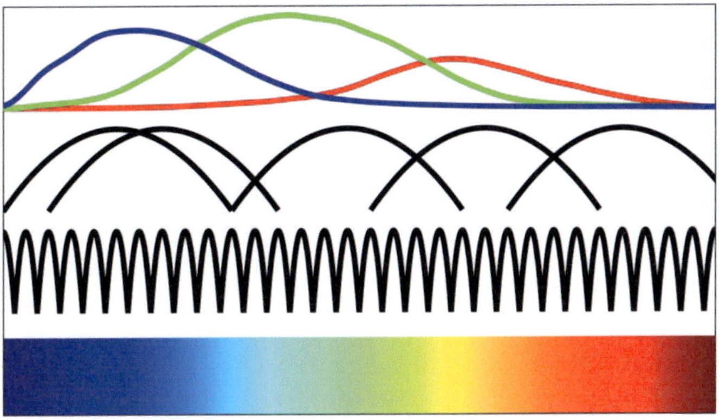

Figure 1.1 Qualitative illustration of RGB (top), multispectral (middle) and hyperspectral (bottom) imaging wavelength bands.

There exist hyperspectral cameras not only for the visible, but also other parts of the electromagnetic spectrum such as the ultraviolet (UV) and the infrared (IR). The resulting hyperspectral image does not only contain color information of the acquired scene, but also spectroscopic information about each image pixel. For this reason, hyperspectral imaging is also referred to as imaging spectroscopy. Figure 1.2 shows an example of a hyperspectral data cube resulting from the acquisition in the near IR.

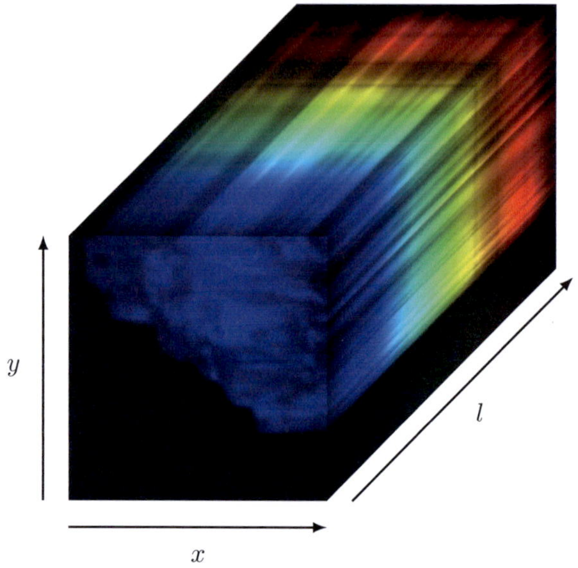

Figure 1.2 Hyperspectral data cube of a mineral sample with the spatial indices x and y and the spectral index l.

The rich spatial and spectral information contained in hyperspectral images allows for accurate retrieval of characteristics of the considered scene. Downsides of hyperspectral imaging, however, are the high hardware effort and cost, as well as the large size of the acquired images which requires both lots of disk space and sophisticated processing algorithms.

Historically, hyperspectral imaging has its roots in remote sensing [49]. Spaceborne and airborne hyperspectral sensors acquire images of the earth's surface or the atmosphere [7]. The analysis of the acquired images allows for a plethora of possible applications such as vegetation monitoring [142].

While it is possible to consider each pixel spectrum of a hyperspectral image individually and, e.g., compare it with a target spectrum for classifying it, there are more sophisticated methods for information retrieval.

These methods can be coarsely split into the following main groups [17]:

- **Classification** performs the automatic assignment of each pixel to one of the available classes using a classifier based on spectral and possibly spatio-spectral features. Using classification in many cases assumes that each pixel spectrum mostly consists of only one (pure) material, but there are also approaches to mixed pixel classification, see for instance [26].
- **Target detection** Due to the large sensing distance in remote sensing, the spatial resolution is very low, i.e., one pixel covers a rectangular area with several or up to dozens of meters edge length on the ground. This leads to linearly mixed pixels (Fig. 2.1), as one pixel does not only contain, say, e.g., a tree, but also a large area of grass around it. For this reason, the acquired spectrum of this pixel is a superposition of the spectra of tree and grass. So-called target detection, a subpixel method, aims at answering the question if a specific spectrum, say tree, is present in a pixel or not. It basically just aims at delivering this binary decision and does not provide a quantitative answer [25].
- **Spectral unmixing**, in contrast to target detection, denotes the process of fully decomposing all acquired pixel spectra into the pure spectra and their spatial fractions, the so-called abundances. Figure 1.3 illustrates this decomposition. Spectral unmixing is the main scope of this thesis.

Further development as well as decreasing costs for hyperspectral imaging hardware has recently lead to a more widespread use of hyperspectral imaging in laboratory and industrial applications. Examples include medical diagnosis [86] and food safety [43, 137]. Hyperspectral imaging is also widely used in chemometrics [3, 125].

Although there are still challenges in terms of sensor size, cost and data analysis, it is conceivable that spectral sensors will be integrated into smartphones in the next couple of years. Research organizations such as VTT Technical Research Centre of Finland Ltd [147] and companies such as Consumer Physics [32] recently have claimed that they have made significant progress in miniaturization and data processing.

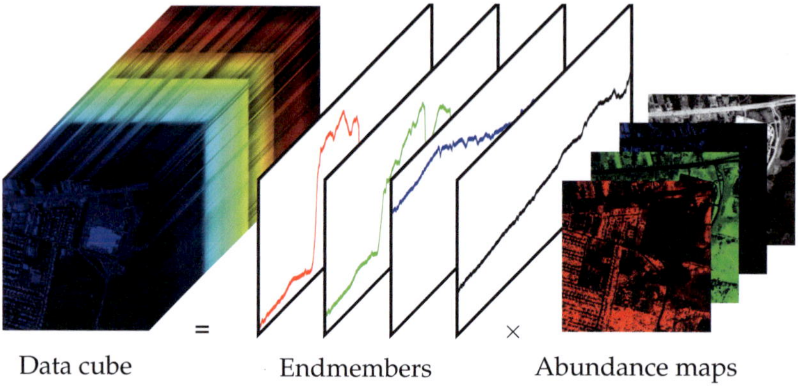

Figure 1.3 Spectral unmixing goal: data cube decomposition into pure spectra (endmembers) and their spatial abundances [24].

1.2 Scope of the Thesis and Own Contribution

The main goal of this thesis is to enhance both the quality and calculation speed of spectral unmixing by incorporating spatial and spectral information. The newly developed methods can be applied to images from both remote sensing and ground applications. An application of particular interest for this thesis is industrial sorting. One secondary objective is therefore the application of unmixing methods to industrial sorting scenarios.

Many sorting machines are using the same principle: a linescan camera is placed over a conveyor belt. The camera image is analyzed and a sorting decision is made, i.e., it is decided if a specific object is to be removed from the stream or not. At the end of the conveyor belt, air nozzles blow out the undesired objects. The sorting principle is depicted in Fig. 1.4.

Conventional RGB sensors are not able to distinguish between objects that have the same color but are of different material, such as small white stones and white plastic particles. At this point, hyperspectral cameras come into play. The acquired pixel spectra are more discriminative than a simple RGB tuple. Sensors using the pushbroom principle are especially suited; they acquire the full spectra of a pixel line in one shot. Making use of the spatial motion of the conveyor belt and sequentially acquiring

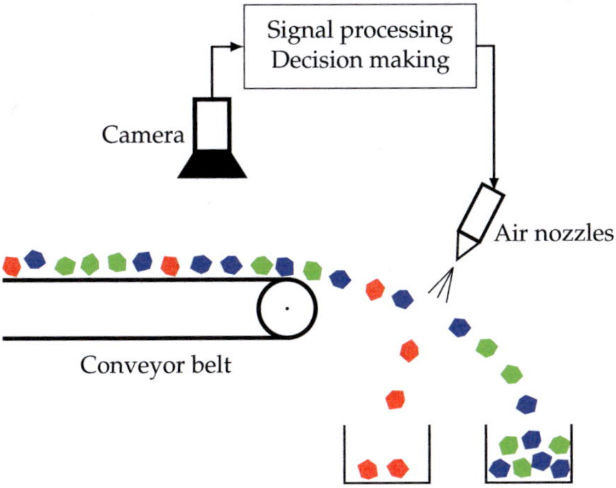

Figure 1.4 Belt sorter principle.

line by line builds a full 3D data cube. The challenges limiting the more widespread use of hyperspectral cameras for such purposes are the general issues such as high cost and the high data volume that needs to be processed and analyzed. Furthermore, hyperspectral sensors have quite low spatial resolution in comparison with RGB sensors. This is due to the fact that the available sensor chip area has to record not only the full spatial, but also spectral information. As this area is limited, there is always a balance between spectral and spatial resolution. The low spatial resolution in combination with high belt speeds leads to the same situation as in the remote sensing scenario: the objects on the belt are blurred, and each pixel most likely contains a mixture of different materials.

While the main focus of this thesis is set on the development of unmixing methods that work with reflectance images, a measurement setup for acquiring hyperspectral fluorescence images has been established during the work on this thesis, too. The developed denoising algorithm explained in Sec. 4.7 is applied to the acquired fluorescence images with the results presented in Sec. 5.3. The classification results of mineral samples based on hyperspectral fluorescence images are published in [165, 167–170, 172].

1.3 Thesis Overview

The present thesis is split into six chapters. Following the introduction in this chapter, the fundamentals required for the remainder of the thesis will be explained in Chapter 2. Chapter 3 will provide an overview of the most important hyperspectral image processing methods presented in the literature so far. The focus of this literature review is on methods that are within the context of the newly developed methods. The first major part of the thesis is the description of the newly developed methods in Chapter 4.

First, a newly developed denoising algorithm will be presented, followed by unmixing methods incorporating spatial abundance derivatives. Afterwards, the unmixing of clustered images and using image pyramids during unmixing will be explained. While these two methods are based on the spatial neighborhood and possibly spectral similarity of pixels, another new method performs a 3D discrete cosine transform (DCT), therefore exploiting both spatial and spectral adjacency at the same time. This method opens the field for many new extensions. First, most images are represented sparsely in the 3D DCT domain, so by discarding the smallest coefficients, the image can be approximated with very low error. Second, it is possible to use inverse matrices, leading to an optimization problem with convex data fidelity term. This is contrasting the vast majority of existing unmixing algorithms, which use the non-convex conventional data fidelity term explained in Sec. 3.2. The third and maybe most important point is how the signal-to-noise ratio (SNR) depends on the coefficient position in the transformed cube. The scattering of the signal energy and therefore the SNR in the cube can be influenced by scrambling regions of the original image. The unmixing then only uses the regions with the highest SNR. Alternatively, orthogonal transforms other than the DCT that suitably shape the SNR could be designed specifically. One example is the signal-dependent singular value decomposition (SVD) which is also used for unmixing in a 3D manner.

The second major part of the thesis in Chapter 5 is dedicated to the evaluation of the presented newly developed methods. Using hyperspectral images from both remote sensing and laboratory environment, all methods will be compared. The last Chapter 6 will conclude the thesis and give an outlook on possible future research directions.

2 Physical Foundations and Mathematical Basics

The main goal of this chapter is to provide an overview of the laws of physics and mathematical methods that are helpful for understanding the remainder of the thesis. For fulfilling this goal while maintaining conciseness, only a brief explanation of each subtopic and references for further reading will be given.

2.1 Spectral Imaging

While the human eye is only able to perceive electromagnetic radiation in the visible wavelength range from about 380 to 780 nm [12], the acquisition of hyperspectral images is not limited to this wavelength range. There are commercially available sensors that operate in other parts of the electromagnetic spectrum. A possible subdivision of the electromagnetic spectrum in the considered wavelength range where hyperspectral imaging typically operates is given in Tab. 2.1.

The vast majority of hyperspectral imaging applications use the frequency range from the visible (VIS) to the short-wavelength infrared (SWIR). Hyperspectral analysis in the visible wavelength range mainly resolves the color more accurately (intensity over wavelength), which, e.g., allows for the discrimination of different types of chlorophyll. For instance for food analysis applications, the VIS and the NIR (near-infrared) up to about 1000 nm is the most often used wavelength range [137].

The reason why NIR is frequently used can be explained with physics. The distinctive spectral features of substances are mainly due to molecular vibrations. The two general types of such vibrations are stretching and bending. The frequency (or wavelength) of a molecular vibration is determined by the energy difference between two vibrational states

[135]. Due to quantum physics, only discrete levels of energy are allowed, which has the consequence that only specific, discrete wavelengths are observed. Molecular fundamental transitions can be mainly found in the mid-wavelength infrared (MWIR) region. The NIR exhibits absorption transitions of overtones and combination tones of these fundamental transitions. For this reason, this wavelength range is used for both qualitative and quantitative analysis of components in analytical chemistry [138].

For the above mentioned reasons, remote sensing devices operate very often in both the visible (VIS) and near-infrared (NIR) wavelength region. The well-known Airborne Visible/Infrared Imaging Spectrometer (AVIRIS) [4] which provided lots of data often used in the remote sensing community has 224 contiguous bands with wavelengths from 400 to 2500 nm. The Hyperspectral Digital Imagery Collection Experiment (HYDICE) operates in the same range [99]. The EnMAP hyperspectral imager [44] uses a slightly different range from 420 to 2450 nm.

Table 2.1 Wavelength regions. The abbreviations in parentheses denote the region name in the DIN 5031-7 [37].

Region name	Abbreviation	Wavelength/nm
Ultraviolet	UV-B (UV-B)	280 – 315
Ultraviolet	UV-A (UV-A)	315 – 380
Visible	VIS	380 – 780
Near-infrared	NIR (IR-A)	780 – 1400
Short-wavelength infrared	SWIR (IR-B)	1400 – 3000
Mid-wavelength infrared	MWIR (IR-C)	3000 – 8000
Long-wavelength infrared	LWIR (IR-C)	8000 – 15000
Far infrared	FIR (FIR)	$15000 - 10^6$

2.1.1 Spectral Mixing Models

The measured pixel spectra in hyperspectral images are in many cases mixtures of the spectra of pure substances, the so-called endmembers [73]. If the spatial resolution of the imaging sensor is low compared to the size of the analyzed objects, the mixing can be described by the linear mixing

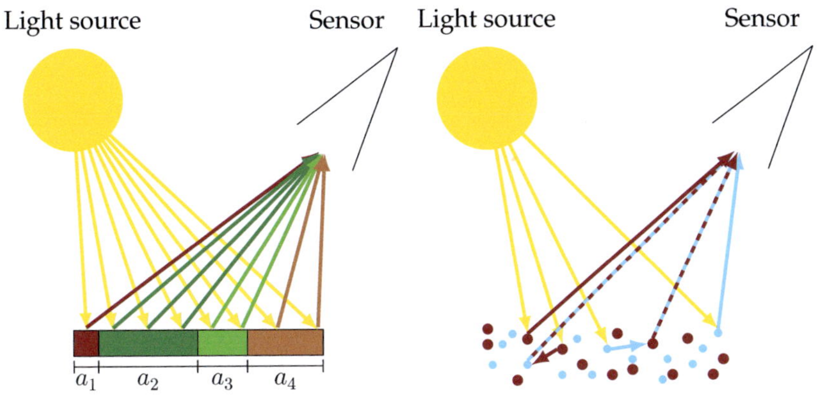

(a) Linear mixing. **(b)** Nonlinear mixing.

Figure 2.1 Spectral mixing models [17]. Linearly mixed spectra are the endmember spectra weighted by the proportions of their surface cover. Nonlinearly mixed spectra arise from light interacting with particles of different materials.

model (LMM) with sufficient accuracy. The effect is illustrated in Fig. 2.1(a) and causes the measured pixel spectra to be endmember spectra weighted by the proportions of their surface cover. These proportions are called abundances. The assumption that the spatial resolution is low is mostly valid in remote sensing, but also hyperspectral cameras used for laboratory and industrial applications have low spatial resolution. This is due to the fact that they have to record both spatial and spectral information on the same sensor area which requires balancing between spatial and spectral resolution. Images acquired by current hyperspectral cameras typically consist of several hundred pixels along each of the two spatial dimensions. Also, one needs to consider that hyperspectral images require a large amount of disk space. Because of the high number of spectral bands, hyperspectral images easily have a size of several hundred megabytes. For this reason, one limits the spatial number of pixels. It can be expected that further development will increase both spatial and spectral resolution in the future.

Considering one pixel, the linear mixing model [73] is given by

$$\mathbf{y} = \mathbf{M}\mathbf{a} + \boldsymbol{\psi}\,. \tag{2.1}$$

The measured spectrum $\mathbf{y} \in \mathbb{R}^{L\times 1}$ is a linear combination of all endmembers contained in the scene. These are collected in the columns $\mathbf{m}_r = [m_{1,r}, \ldots, m_{L,r}]^{\mathrm{T}}$ of the endmember matrix $\mathbf{M} \in \mathbb{R}^{L\times R}$. L denotes the number of bands of the hyperspectral image, and R represents the number of endmembers that are present in the scene. The endmember spectra are weighted by the abundance vector $\mathbf{a} \in \mathbb{R}^{R\times 1}$ of the given pixel; this vector encodes the abundances $[a_1, \ldots, a_R]^{\mathrm{T}}$ to which each of the endmember spectra contributes to the mixed pixel $\mathbf{y}$. The vector $\boldsymbol{\psi}$ accounts for noise as well as model errors.

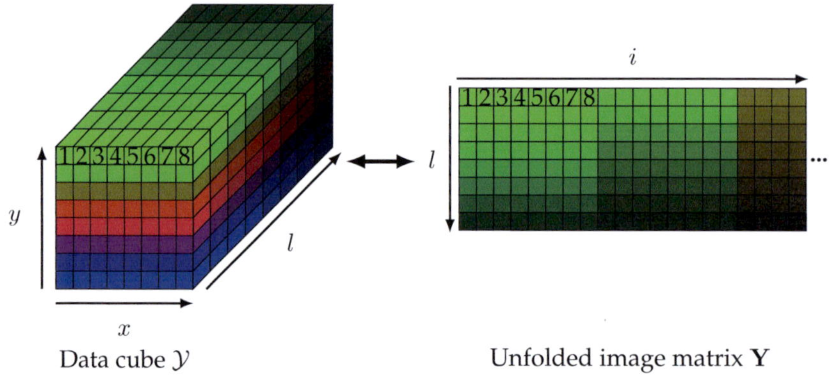

Figure 2.2 Unfolding of the data cube $\mathcal{Y}$ to the image matrix $\mathbf{Y}$. The spectra of the first row of pixels along x, denoted by the numbers 1–8, are put in the first 8 columns of the image matrix. The second row of pixels follows in the next columns etc. The index i denotes the pixel spectra in the columns of $\mathbf{Y}$.

Considering the full 3D hyperspectral data cube $\mathcal{Y} \in \mathbb{R}^{X\times Y\times L}$ with $X \times Y$ spatial pixels and L bands, its unfolded version $\mathbf{Y} \in \mathbb{R}^{L\times I}$ contains all pixel spectra in its columns. This conversion is illustrated in Fig. 2.2. The number $I = X \cdot Y$ denotes the total number of spatial image pixels. For the whole image, the LMM (2.1) is given by

$$\mathbf{Y} = \mathbf{M}\mathbf{A} + \boldsymbol{\Psi} . \tag{2.2}$$

Just like the image matrix $\mathbf{Y}$ contains the pixel spectra in its columns, the abundance matrix $\mathbf{A} \in \mathbb{R}^{R\times I}$ contains the respective pixel abundance vectors in its columns. For preventing the notation within the respective application from becoming bulky, an ambivalent convention is defined, and for each application, the most suitable version of it is used. This affects the hyperspectral image itself, but mainly the calculated abundances. Just like the hyperspectral image is the tensor $\mathcal{Y} \in \mathbb{R}^{X \times Y \times L}$, the abundances of the R endmembers at each of the $X \times Y$ pixels form the 3D tensor $\mathcal{A} \in \mathbb{R}^{X \times Y \times R}$ (Fig. 2.3). In most cases, the unfolded abundance matrix $\mathbf{A}$ will be used. It contains the abundances of all pixels, the vectors $\mathbf{a}_i$, $i = 1, \ldots, I$, in its columns. Especially for spatial regularization (see Sec. 4.1), however, it is necessary to consider the abundance map $\mathbf{A}_r \in \mathbb{R}^{X \times Y}$ of endmember r, i.e., the r-th row of the unfolded matrix $\mathbf{A}$ rearranged to the original image size $X \times Y$. When it is necessary to address the abundance of endmember r at pixel (x, y) in this spatial notation, this is done by $\mathbf{A}_r(x, y)$. The same element can be found in the unfolded matrix $\mathbf{A}$ in row r at column i. The abundance representations with the unfolded matrix $\mathbf{A}$ and the spatial abundance maps $\mathbf{A}_r(x, y)$, $r = 1, \ldots, R$, respectively, can be converted from one to the other and vice versa.

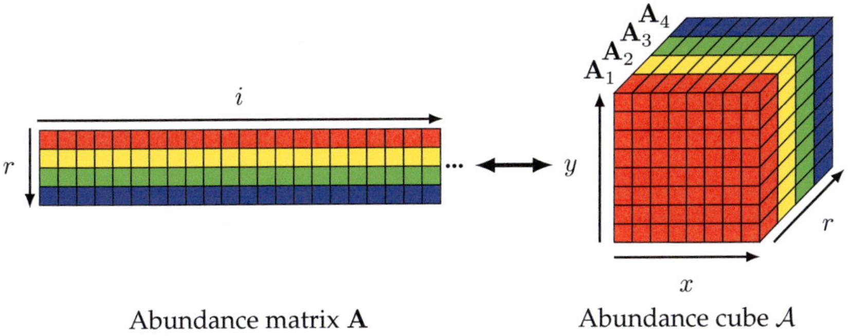

Figure 2.3 Folding the abundance matrix $\mathbf{A}$ to the tensor $\mathcal{A}$ whose slices are the matrices $\mathbf{A}_r$.

In cases in which light interacts with more than one pure substance before detection (Fig. 2.1(b)), the LMM is not accurate enough. In this case, nonlinear mixing models (NMMs) are used. The probably most prominent one is the Hapke model [55], but there are also other, more recent models [38]. Nonlinear mixing occurs mainly in situations where there is high spatial resolution, but also when substances are mixed on a molecular level. Even in such cases, however, the linear mixing model can sometimes be applied with sufficient accuracy, see especially chemical analyses such as, e.g., [112].

For taking physical plausibility into account, the following two constraints will be incorporated in all unmixing methods:

- All elements of the endmember and abundance matrices $\mathbf{M}$ and $\mathbf{A}$ must be nonnegative (nonnegativity (NN) constraint)

$$m_{l,r} \geq 0 \quad \forall l, r \tag{2.3}$$

$$a_{r,i} \geq 0 \quad \forall r, i\,. \tag{2.4}$$

- The sum of all abundances of each pixel has to sum to one:

$$\sum_{r=1}^{R} a_{r,i} = 1 \quad \forall i\,. \tag{2.5}$$

  This constraint is called the sum-to-one constraint (STO constraint). Its application is reasonable in case the endmembers comprise all the pure substances present in a scene.

2.1.2 Spectral Variability

In an ideal world, the endmember spectra of each substance are constant over time and space. In reality, however, there are various effects leading to spatially and temporally varying endmember spectra. In remote sensing, these are atmospheric conditions, but also environmental effects such as the seasons that influence the appearance of vegetation. One very important factor that does not only influence the earth surface, but also laboratory and industrial scenes is illumination. In remote sensing, one major influencing factor is the sunlight; depending on, e.g., solar elevation, solar azimuth, and local incidence angle [157], the endmember shape varies with time and spatial position. For ground applications, the object geometry as well as surface roughness and object reflection properties have a large impact on the acquired spectra. Minerals exhibit spectral variability depending on grain size and texture [157]. Furthermore, objects may show intrinsic spectral variability. For two laboratory images, Fig. 2.4 illustrates the spectral variability.

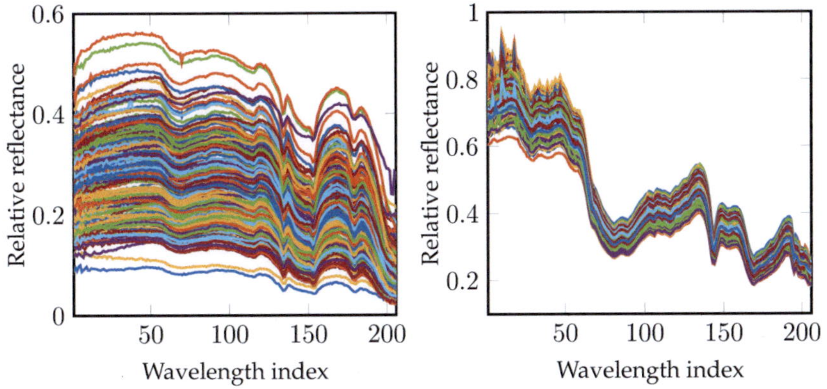

(a) Pixel spectra of the mineral sample shown in Fig. 1.2.

(b) Pixel spectra of one pure powder compartment of the White Powders image used in [176] and described in detail in Sec. 5.1.1.

Figure 2.4 Examples of spectral variability.

Incorporating spectral variability into the unmixing process can greatly improve the unmixing quality. Practically, there are two major options for doing so: Zare et al. [157] mention endmember sets (i.e., there are multiple spectra of differing shape considered for one endmember), and modelling endmembers with statistical distributions. Specific methods that take care of the problem are described in [133, 157], one of the easiest being the consideration of only the band regions that exhibit the lowest variability. Throughout this thesis, endmember variability will not be incorporated in the methods, but there is an awareness of its existence and influence.

2.2 Mathematical Background

This section explains vector and matrix norms, as well as the fundamentals of optimization that are required for this thesis. It closes with probability densities and their scaling.

2.2.1 Vector Norms

Let $\mathbf{a} \in \mathbb{R}^R$. A frequently used family of vector norms are the ℓ_p norms with $p \geq 1$ defined by [19, Sec. A.1.3]

$$\|\mathbf{a}\|_p = \left(\sum_{r=1}^{R} |a_r|^p \right)^{1/p} . \tag{2.6}$$

Its most prominent members are the Euclidean norm for $p = 2$

$$\|\mathbf{a}\|_2 = \sqrt{\sum_{r=1}^{R} a_r^2} \tag{2.7}$$

and the ℓ_1 norm for $p = 1$

$$\|\mathbf{a}\|_1 = \sum_{r=1}^{R} |a_r| . \tag{2.8}$$

The maximum norm ℓ_∞ is defined by

$$\|\mathbf{a}\|_\infty = \max(a_1, \ldots, a_R) . \tag{2.9}$$

One very important measure which is not a norm, not even a quasinorm, but plays a very important role in sparse coding and compressive sensing is the ℓ_0 measure [35]: it simply counts the number of nonzero elements of $\mathbf{a}$. Its relevance lies in minimizing the number of coefficients required for sparse approximation of a signal. Because its use results in a combinatorial problem which is NP-hard (non-deterministic polynomial-time hard), it is often relaxed by the ℓ_1 norm, e.g., in basis pursuit (BP) [29].

2.2.2 Matrix Norms

It is possible to write all matrix elements as a long vector and use a vector ℓ_p norm. Equivalently, the matrix p-norms can be used. For $\mathbf{A} \in \mathbb{R}^{R \times I}$, these norms are calculated using

$$\|\mathbf{A}\|_p = \left(\sum_{r=1}^{R} \sum_{i=1}^{I} |a_{ri}|^p \right)^{1/p} . \tag{2.10}$$

Choosing $p = 2$ yields the Frobenius norm

$$\|\mathbf{A}\|_F = \sqrt{\sum_{r=1}^{R} \sum_{i=1}^{I} a_{ri}^2} \,. \tag{2.11}$$

By contrast, the spectral norm of $\mathbf{A}$ does not have a vector norm equivalent. It is defined by the singular values of $\mathbf{A}$, specifically, it is the maximum singular value:

$$\|\mathbf{A}\|_2 = \sigma_1 \,. \tag{2.12}$$

The spectral norm will be used for extensions of the 3D DCT unmixing procedure discussed in Sec. 4.4.7.

When applied to minimizing the error between model and measured data, the Frobenius norm is equivalent to the least-squares estimator. In turn, the least-squares estimator is the best linear unbiased estimator when the noise of the measurements is uncorrelated and described by a normal distribution with zero mean and equal variance for all measurements [13]. This means that the estimator is unbiased and provides the least variance in its estimations among all linear unbiased estimators. For this reason, the Frobenius norm is often chosen as objective function when estimating model parameters. If the error has different statistics, different metrics/distances have to be used [31].

2.2.3 Spectral Angle

For expressing the similarity between a given spectrum $\mathbf{y}$ and a reference spectrum $\mathbf{z}$, the spectral angle (SA) [75] is an often used measure. It is calculated by

$$\mathrm{SA}(\mathbf{y}, \mathbf{z}) = \arccos\left(\frac{\mathbf{y} \cdot \mathbf{z}}{\|\mathbf{y}\|_2 \cdot \|\mathbf{z}\|_2}\right) = \arccos\left(\frac{\sum_{l=1}^{L} y_l z_l}{\sqrt{\sum_{l=1}^{L} y_l^2}\sqrt{\sum_{l=1}^{L} z_l^2}}\right) , \tag{2.13}$$

where the operation $\cdot$ applied to two vectors denotes the dot product. The spectral angle is therefore the angle between two spectra in the L-dimensional space; its great advantage is that it is independent of multiplicative effects arising for instance from illumination variations.

2.2.4 Optimization Theory

The vast majority of spectral unmixing methods is based on the matrix version of the LMM (2.2). The goal is to determine both the endmember matrix $\mathbf{M}$ and the abundance matrix $\mathbf{A}$ from the given image matrix $\mathbf{Y}$. Based on this model, most unmixing methods use optimization to perform this decomposition. For this reason, some basic concepts of optimization will be briefly recapitulated in the following.

Let the domain $\mathrm{dom}\, Q$ of the function $Q : \mathbb{R}^R \to \mathbb{R}$ be a convex set. If for all $\mathbf{a}, \mathbf{b} \in \mathrm{dom}\, Q$, $0 \leq \theta \leq 1$,

$$Q(\theta\mathbf{a} + (1-\theta)\mathbf{b}) \leq \theta Q(\mathbf{a}) + (1-\theta)Q(\mathbf{b}) \tag{2.14}$$

holds, Q is convex [19]. The geometric interpretation of this relation is given in Fig. 2.5: the connecting line between all possible points $(\mathbf{a}, Q(\mathbf{a}))$ and $(\mathbf{b}, Q(\mathbf{b}))$ is always above the function. Without going into detail about the existence of stationary points and other function properties (see [19]), convexity is an advantageous property for a function that is to be minimized: it will only have one minimum and slope towards this minimum from any given point. In this favorable case, gradient descent methods can be used for minimizing such functions and finding the global minimum.

While there are many operations that preserve convexity [19], the most important one in the context of this thesis is that the sum of convex functions also is a convex function.

The minimization problem

$$\begin{aligned} &\text{minimize } Q_0(\mathbf{a}) \\ &\text{subject to } Q_g(\mathbf{a}) \leq 0,\ g = 1, \ldots, G, \\ &\text{subject to } Q_h(\mathbf{a}) = 0,\ h = G+1, \ldots, H \end{aligned} \tag{2.15}$$

is called convex if [19]

- the objective function $Q_0(\mathbf{a})$ is convex and
- the inequality constraint functions $Q_g(\mathbf{a})$, $g = 1, \ldots, G$ are convex and
- the equality constraint functions $Q_h(\mathbf{a})$, $h = G+1, \ldots, H$ are affine.

Each local minimum of the minimization problem (2.15) is a global minimum.

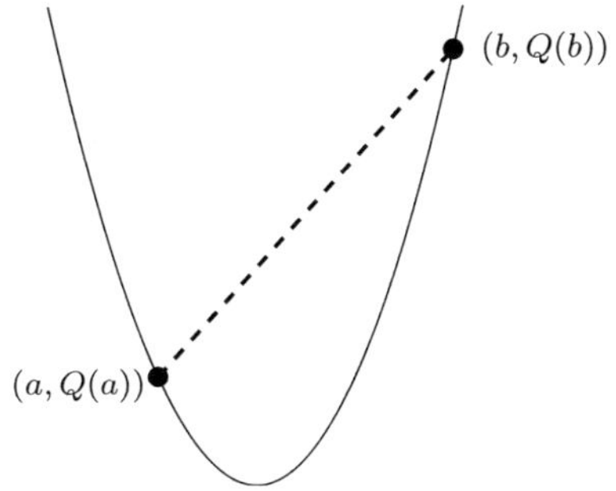

Figure 2.5 Convex function in the one-dimensional case: the connecting line between $(a, Q(a))$ and $(b, Q(b))$ is always above the function.

Finding such a minimum with gradient descent methods requires the calculation of the objective function gradient. Even if the objective function depends on a matrix, say $\mathbf{A} \in \mathbb{R}^{R\times I}$, it can be converted to a function depending on the vector $\mathrm{vec}(\mathbf{A}) \in \mathbb{R}^{R\cdot I\times 1}$ which consists of all elements of matrix $\mathbf{A}$ stacked on top of each other, as long as matrix norms such as the p-norms (2.10) are used. However, when sticking with matrices, finding the gradient of a function $Q(\mathbf{A}) : \mathbb{R}^{R\times I} \to \mathbb{R}$ consists in deriving the function by each element a_{ri} of $\mathbf{A}$ and arranging the results to the dimensions of $\mathbf{A}$ in the same order as the matrix elements. Useful calculation rules for derivatives of a scalar function by a matrix are given for instance in [111].

The function $Q(\mathbf{M}, \mathbf{A}) : \mathbb{R}^{L\times R} \times \mathbb{R}^{R\times I} \to \mathbb{R}$ is called biconvex if it is convex in $\mathbf{M}$ for constant $\mathbf{A}$ and vice versa. Note that for such a function in general, a local minimum is not a global minimum.

2.2.5 Statistics Basics

The probability density of a random variable z is denoted by $f_{\mathrm{z}}(z)$. Scaling the variable by the constant $k \in \mathbb{R}\backslash\{0\}$, resulting in the new variable

$$\mathrm{w} = k \cdot \mathrm{z}\,, \tag{2.16}$$

means that the density converts to

$$f_{\mathrm{w}}(w) = 1/\left|k\right| f_{\mathrm{z}}(z/k)\,. \tag{2.17}$$

A normally distributed random variable z has the density

$$f_{\mathrm{z}}(z) = \frac{1}{\sigma\sqrt{2\pi}} \mathrm{e}^{-\frac{z^2}{2\sigma^2}}\,. \tag{2.18}$$

The probability density of a sum of random variables is the convolution of their densities. As a consequence, the inversely Fourier transformed densities (i.e., the characteristic functions [146]), can be multiplied.

2.3 Poisson Noise and EMCCD Noise Characteristics

In the following, a hyperspectral fluorescence imaging setup created during the work on this thesis is briefly described. A newly developed multidimensional denoising algorithm described in Sec. 4.7 is applied to the images acquired with the setup. In this setup, spectral filtering is performed with an acousto-optical tunable filter (AOTF, *Gooch&Housego HSi-300*). This filter can be tuned to a specific wavelength and bandwidth. At each of these channels, a full grayscale image is acquired (so-called spectral scanning). The applied spectral resolution is 4 nm in the wavelength range from 450 nm to 790 nm. In the present section, the noise characteristics of the used EMCCD (electron-multiplying charge-coupled device) camera will be explained. As it can be approximated by a Poisson distribution that is found quite often in imaging in general, one version of the denoising algorithm was designed specifically for this type of noise. The other version accounts for normally distributed noise and will be applied to remotely sensed images, see Sec. 5.3.1.

Due to the discrete nature of light and stochastic fluctuations [12, 71], the number of photon arrivals at the imaging sensor during the exposure time Δt is not the same when repeating the acquisition. For this physical reason, the number of incident photons $\mathtt{i}$ arriving at the sensor during Δt has to be described by a random variable, even if no noise would be added by the sensor and readout electronics [71, Sec. 3.4.1]. As a consequence, the obtained pixel value $\mathtt{g}$ is also a random variable. The number of incident photons $\mathtt{i}$ has a Poisson PDF:

$$f_{\mathtt{i}}(i) = \frac{\lambda_{\mathtt{i}}^{i}}{i!} \mathrm{e}^{-\lambda_{\mathtt{i}}}, \tag{2.19}$$

where $\lambda_{\mathtt{i}}$ equals the expected value $\mathrm{E}\{\mathtt{i}\}$. It is important to note that for the Poisson PDF, its expected value equals its variance: $\mathrm{E}\{\mathtt{i}\} = \mathrm{Var}\{\mathtt{i}\} = \lambda_{\mathtt{i}}$, which means that this kind of noise is signal dependent. For large values of $\mathtt{i}$, the Poisson distribution can be approximated quite well by a normal distribution [100, Sec. 2.1]. Not every incident photon generates a photoelectron in the sensor. The expected value $\lambda_{\mathtt{n}}$ of the number of photoelectrons $\mathtt{n}$ generated within the time interval Δt is therefore given by [12]

$$\lambda_{\mathtt{n}} = \lambda_{\mathtt{i}} \alpha \,. \tag{2.20}$$

The quantum efficiency α denotes the fraction of photons that are absorbed and generate a photoelectron. In addition to this natural noise component, there is the noise added to the signals in the imaging sensor. According to the EMVA 1288 standard [45] for industrial cameras, the noise added to the digital signal of each camera pixel has the following three major components which are added using a linear signal model:

- signal dependent shot noise,
- signal independent normally distributed sensor read out and amplification noise,
- uniformly distributed quantization noise.

In addition, there is the Poisson distributed, temperature dependent dark current whose mean value is considered as well.

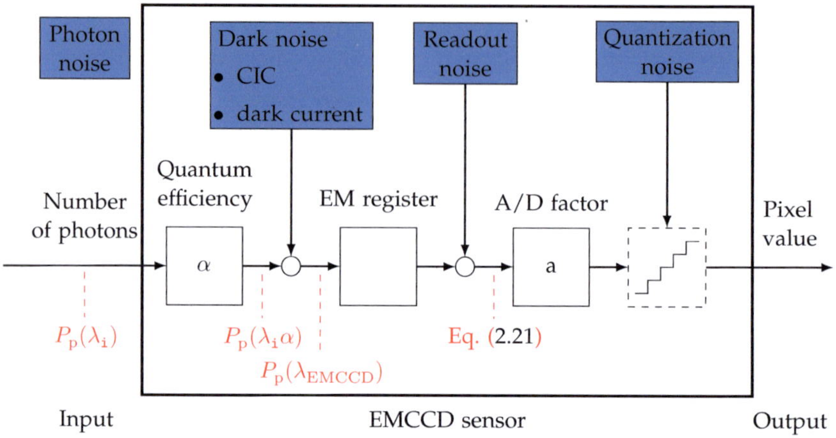

Figure 2.6 Mathematical EMCCD model. The main difference to a conventional CCD sensor is the EM register block. Red font describes the probability density at the given stage; $P_p(\lambda)$ denotes the Poisson distribution. The quantization noise is neglected in case of high EM gain, see text.

Extending conventional CCD image sensors (see, e.g., [71] for a description), EMCCDs have an additional electron multiplication register amplifying the signal before readout, see Fig. 2.6. Therefore, the effective readout noise is greatly reduced; however, the multiplication register is an additional noise source that needs to be considered and modelled. For this reason, a noise model different from the EMVA 1288 standard is considered. The following description of the noise characteristics of EMCCDs is based on [60]. The notation has been adapted for the sake of consistency within this thesis. For EMCCDs, the PDF for obtaining the pixel value g is given by [60]

$$f_g(g) = \frac{1}{\sqrt{2\pi}\sigma} \exp\left(-\lambda_{EMCCD} - \frac{(ag)^2}{2\sigma^2}\right) + \frac{2}{G} f_\chi(2\lambda_{EMCCD}; 4, 2ag/G) . \tag{2.21}$$

Only the case $g > 0$ is considered here; due to the noise characteristics, negative values are also possible and occur at very low signal intensities. In (2.21), a is the analog-to-digital proportionality factor which stands for the number of electrons per image value, while G is the gain of the multiplication register. In accordance with EMVA 1288, the first term of (2.21) represents the readout noise described by a normal distribution with standard deviation σ. This term is weighted by $\exp\left(-\lambda_{\mathrm{EMCCD}}\right)$. The second term is not present in the EMVA 1288 standard and models the amplified signal. The electrons go through a number of register stages. In each stage, the electrons are likely to produce additional electrons by ionization with a certain probability. In the end, there is a large number of electrons arising from just a few photoelectrons. Describing the process mathematically and applying some approximations (see [60] for the exact derivation), the second summand of (2.21) results. f_{χ} is the non-central χ^2 distribution [123, Sec. 5.3] for $2\lambda_{\mathrm{EMCCD}}$. Its arguments are 4 degrees of freedom and the noncentrality parameter $2ag/G$. The quantization noise in EMCCDs can be neglected due to the high gain values, meaning that one quantization step would account only for a fraction of an electron before the multiplication process.

The expected number of electrons λ_{EMCCD} before the multiplication register amounts to

$$\lambda_{\mathrm{EMCCD}} = \lambda_{\mathtt{i}} \cdot \alpha + d \cdot \Delta t + c\,, \tag{2.22}$$

where d stands for the dark current rate in electrons per pixel per second. The variable c denotes the clock induced charge (CIC) in electrons, i. e., additional electrons that are generated during the shift of the electrons through the pixels on the chip for readout. Due to the fact that the camera chip is cooled to about −85 °C, the temperature-dependent dark current is considered negligible.

Considering equations (21)–(33) from [60] (see also the appendix of [194] for a detailed proof), the expected value of the probability density (2.21) can be calculated by

$$\mathrm{E}\{\mathrm{g}\} = \frac{G}{a}\lambda_{\mathrm{EMCCD}} \tag{2.23}$$

and the variance amounts to

$$\mathrm{Var}\{\mathrm{g}\} = \mathrm{e}^{-\lambda_{\mathrm{EMCCD}}}\frac{\sigma^2}{a^2} + 2\lambda_{\mathrm{EMCCD}}\frac{G^2}{a^2}\,. \tag{2.24}$$

From (2.23) and (2.24), the ratio between variance and mean is

$$\frac{\mathrm{Var}\{\mathrm{g}\}}{\mathrm{E}\{\mathrm{g}\}} = \frac{1}{a}\left(\mathrm{e}^{-\lambda_{\mathrm{EMCCD}}}\frac{\sigma^2}{G\lambda_{\mathrm{EMCCD}}} + 2G\right). \tag{2.25}$$

The first term will be neglected in the following: for the specific camera in the setup with the selected gain $G = 300$ and considering only one single electron ($\lambda_{\mathrm{EMCCD}} = 1$), the first term is much smaller than the second one. For increasing λ_{EMCCD}, this ratio becomes even larger. Therefore, the approximation is

$$\frac{\mathrm{Var}\{\mathrm{g}\}}{\mathrm{E}\{\mathrm{g}\}} \approx \frac{2G}{a}. \tag{2.26}$$

For the Poisson distribution, the variance equals the mean. By multiplying the generic random variable x with a constant value k, the mean and the variance of this variable are altered according to

$$\mathrm{Var}\{k\mathrm{x}\} = k^2\mathrm{Var}(\mathrm{x}) \tag{2.27}$$

and

$$\mathrm{E}\{k\mathrm{x}\} = k\mathrm{E}(\mathrm{x}). \tag{2.28}$$

Choosing the multiplication factor $k = \frac{a}{2G}$, the ratio (2.26) becomes one, i.e., expected value and variance become equal. This means that by multiplying the obtained image values by k, the resulting random variable has a variance that equals its mean. In this case, it is possible to compare the EMCCD density (2.21) with a regular Poisson density: By setting λ_{EMCCD} to fixed values $\lambda_{\mathrm{EMCCD},1} = 5$, $\lambda_{\mathrm{EMCCD},2} = 15$, it is shown graphically in Fig. 2.7 that the Poisson density is a very good approximation of (2.21) multiplied by k. Note that the chosen values of λ_{EMCCD} are within the range of the obtained fluorescence images, see Sec. 5.1.2. Sometimes, as little as 5 photons per pixel have been registered. Even in such case, the dominant term in (2.22) is the photoelectrons; as mentioned before, the dark current is negligible.

The plots in Fig. 2.7 have been obtained with an experimentally measured a: Using (2.23), the expected value of the image values is $\frac{G}{a}$ times

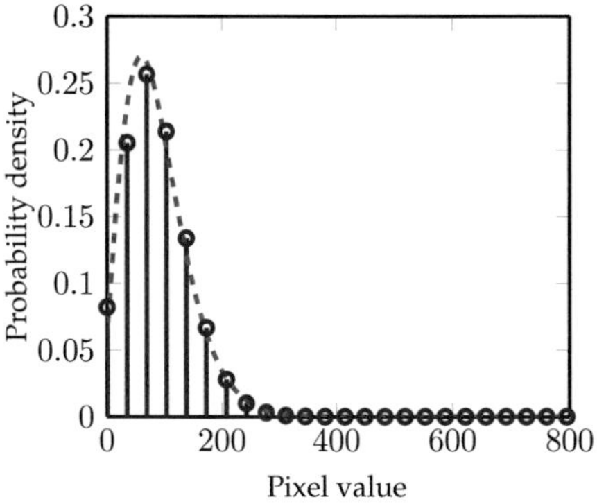

(a) $\lambda_{\mathrm{EMCCD},1} = 5$

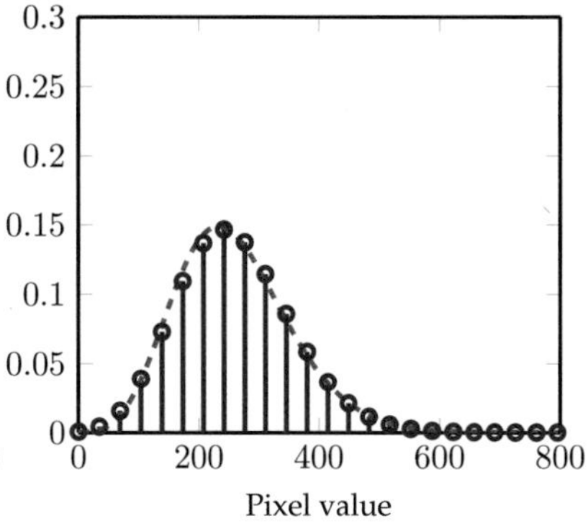

(b) $\lambda_{\mathrm{EMCCD},2} = 15$

Figure 2.7 Comparison between EMCCD density (equation (2.21), dashed red line) and Poisson density (blue stems). The mean squared error between the shown discrete values of EMCCD density and Poisson density amounts to $6.77 \cdot 10^{-5}$ (a) and to $1.03 \cdot 10^{-5}$ (b).

the number of electrons λ_{EMCCD}. The factor $\frac{G}{a}$ has been determined experimentally for the used *Andor iXon*$_3$ *897* EMCCD camera and results to 17.28 image values per photoelectron for $G = 300$ [194].

The previous paragraphs have shown that the noise present in the digital pixel values obtained with an EMCCD sensor can be described by a Poisson distribution with sufficient accuracy. For this reason, the denoising algorithm that will be presented in Sec. 4.7 has a version that is adapted to this kind of noise, see Sec. 4.7.1. When conventional sensors are used that provide images with dominant normally distributed noise, the appropriate algorithm version described in Sec. 4.7.2 should be used.

3 State of the Art in Hyperspectral Image Processing

This chapter provides a short overview on hyperspectral image processing in general and describes the main groups of methods in Sec. 3.1. As the main topic of this thesis is spectral unmixing, the following Section 3.2 takes a closer look at the state of the art regarding this area of research. The last Section 3.3 is concerned with hyperspectral image denoising and explains techniques that have been presented in the literature so far.

3.1 Overview

Hyperspectral image processing can be roughly subdivided into two main groups. The first group of methods deals with image preprocessing such as image registration, image fusion, image restoration (see Sec. 3.3), image segmentation [18, 80] and the like. The second, larger group of methods contains information retrieving methods.

Let us first consider methods that can be assigned to the first group. Image registration [30] aims at transferring different images into one common coordinate system. These images can, for instance, differ in size and/or resolution. Especially in hyperspectral image processing, image registration is very often closely connected with pansharpening. Hyperspectral images commonly have high spectral, but low spatial resolution, whereas monochrome, panchromatic or color images in most cases have high spatial resolution, but inherently low spectral resolution. For this reason, hyperspectral images are often fused, or pansharpened, with such images for obtaining a both spatially and spectrally highly resolved image. Loncan et al. [85] give a thorough overview on pansharpening methods and provide a quantitative comparison. Pansharpening has been combined with spectral unmixing in for example [96], where both a hyperspectral (low spatial, high spectral resolution) and a multispectral (low spectral, high spatial resolution) image are unmixed within a joint mixing model. Hyperspectral image restoration can consist of deblurring [2, 82] or denoising (Sec. 3.3), but also dead or hot sensor pixels or lines and stripes [159] degrade the images and need to be removed before further processing. As hyperspectral images consist of a large number of spectral bands which are in many cases highly correlated and therefore exhibit a large degree of redundancy [81], many hyperspectral image processing methods are concerned with band selection. The goal is to select a low number of bands that are representative for the whole image and contain as much as possible of the full amount of information available in the original image. Reducing the number of bands accelerates and sometimes even improves subsequent analyses such as target detection, endmember extraction and classification (see Sec. 1.1 and the next paragraph). Michelsburg et al. [97] used band selection techniques for designing few narrow-band spectral filters that are optimal for solving specific industrial classification tasks.

The second, larger group of methods contains information retrieving methods based on hyperspectral images [115]. As already stated in the introduction (see Sec. 1.1), the main subgroups of hyperspectral information retrieval are classification, target detection and spectral unmixing. While classification is mainly based on more or less pure pixels, target detection and spectral unmixing consider spectrally mixed pixels. Note that there are also techniques that do not follow this coarse taxonomy, e.g., there are algorithms for mixed pixel classification [26].

Hyperspectral image classification aims at classifying each image pixel into one of the given classes. Such methods mainly use the spectral bands as features and thus classify a pixel using its spectrum [22, 54]. Due to the large number of bands, dimensionality reduction [56], sparsity [30] and kernel classification [22] are very often employed. More recent methods incorporate information of the pixel's neighborhood for creating spatio-spectral features [9, 46] to improve the classification. A good general review on hyperspectral image classification can be found in [23].

In many cases, the image pixels cannot be considered as pure, but are mixtures of several pure substances. Target detection aims at providing a binary answer for each pixel, telling if it belongs to the background or if a target is present in it [104]. As this decision is based on the pixel spectra and the target not necessarily constitutes the full pixel, target detection is a subpixel method. Target detection can be subdivided into anomaly detection and signature-based target detection. Anomaly detection aims at determining if the pixel stands out from the background by analyzing its spectral signature and comparing it with the background pixel spectra. By contrast, signature-based target detection specifically searches for the presence of the target spectrum [104].

While target detection only yields a binary decision if there is an anomaly or if a certain spectrum is present in the image pixels, spectral unmixing aims at providing a quantitative analysis of the image pixels. The next section discusses the state of the art in spectral unmixing.

3.2 Spectral Unmixing

Hyperspectral image unmixing has been briefly discussed in Sec. 1.1. Note especially Fig. 1.3 which gives an illustration of the spectral unmixing decomposition goal into the pure (endmember) spectra and their spatial abundances. Before going into detail on spectral unmixing methods, it is useful to take a look at one possible formal definition of spectral unmixing [16]:

Hyperspectral unmixing refers to any process that separates the pixel spectra from a hyperspectral image into a collection of constituent spectra, or spectral signatures, called endmembers and a set of fractional abundances, one set per pixel. The endmembers are generally assumed to represent the pure materials present in the image and the set of abundances, or simply abundances, at each pixel to represent the percentage of each endmember that is present in the pixel.

While the reference discusses the subtleties of this definition, its basic message is that unmixing consists of determining the endmembers and abundances which together yield the acquired full image, or at least a close approximation (cf. Fig. 1.3). Many thoughts and methods described in this section are based on [16].

The unmixing process can be either supervised or unsupervised, see Fig. 3.1. Supervised unmixing refers to the process of determining the endmembers by a so-called endmember extraction algorithm (EEA) in the first step and subsequently estimating the abundances using the determined endmembers in a second abundance estimation step. Unsupervised unmixing, by contrast, means that both endmembers and abundances are determined in the same step.

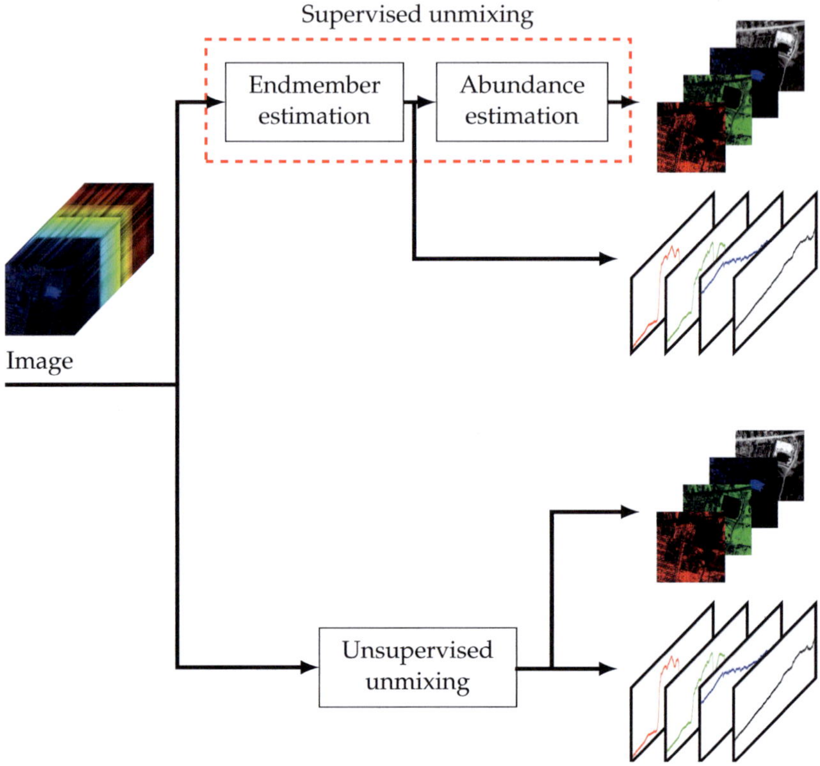

Figure 3.1 Illustration of supervised and unsupervised unmixing.

This thesis is concerned with linear spectral unmixing; however, there are a lot of nonlinear unmixing methods, too. Heylen et al. [59] give an overview of existing nonlinear methods.

Spectral unmixing is in many cases based on the LMM (2.2) in Section 2.1.1. From this linear model, the data fidelity term

$$Q_{\text{data}}(\mathbf{M}, \mathbf{A}) = \|\mathbf{M}\mathbf{A} - \mathbf{Y}\|_{\mathrm{F}}^2 \tag{3.1}$$

is derived. The data fidelity (3.1) has the property of bi-convexity, which means that it is convex in $\mathbf{M}$ for constant $\mathbf{A}$ and vice versa, but not convex in both variables simultaneously. This means that unmixing methods based on this objective function are ill-posed [48] and there is no unique solution. Depending on the initialization of $\mathbf{M}$ and $\mathbf{A}$, very different solutions can result. For this reason, various methods enforcing additional constraints on the matrices $\mathbf{M}$ and $\mathbf{A}$ have been developed. The vast majority of these methods can be expressed by the general scheme

$$Q(\mathbf{M}, \mathbf{A}) = Q_{\text{data}}(\mathbf{M}, \mathbf{A}) + \mu_{\text{spec}} \mathcal{R}_{\text{spec}}(\mathbf{M}) + \mu_{\text{spat}} \mathcal{R}_{\text{spat}}(\mathbf{A}) \,. \quad (3.2)$$

The arguments $(\mathbf{M}, \mathbf{A})$ will be suppressed in all following considerations, because they will be clear from the context. The objective function (3.2) is the one of the general nonnegative matrix factorization (NMF) [31, 77], whose goal is to decompose a given nonnegative matrix into a product of two nonnegative matrices. NMF belongs to the class of methods solving the blind source separation (BSS) problem [31]. Note that not all unmixing algorithms are based on NMF and therefore do not make use of (3.2).

Linear mixing is often based on the geometric concept discovered by Craig [33]: in the R-dimensional space (remember that R stands for the number of endmembers), all pixel spectra are enclosed by the simplex spanned by the endmember spectra. Figure 3.2 depicts this concept. Many endmember extraction algorithms (EEAs) and unmixing methods make use of this concept; some will be discussed in Sec. 3.2.1 and Sec. 3.2.2, respectively. In the case the acquired data do not contain pure pixels of each endmember, the vertices of the simplex are missing. For this scenario, there are algorithms aiming at finding the vertices of the simplex of minimum volume that contains all pixel spectra. If the method is successful, it provides the vertices of this simplex, which correspond to the true endmember spectra. In highly mixed scenarios, the available pixel spectra only fill the inner part of the simplex in Fig. 3.2, and there are not even mixed pixels on the simplex edge. In this case, even minimum volume-based methods fail, and researchers fall back on statistical methods [16, 17]. Such methods often interpret the unmixing problem as a statistical inference problem and use a Bayesian approach. The discussion of such methods goes beyond the scope of this thesis. It should be noted, however, that such statistical approaches allow for the straightforward incorporation of endmember variability, see Sec. 2.1.2.

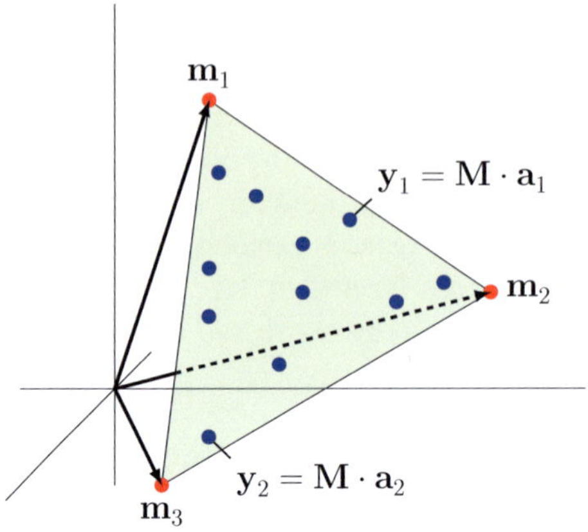

Figure 3.2 Geometric representation of the linear mixing model using the example of three wavelengths and three endmembers. The pixel values in each of the three wavelength channels are plotted along the three axes. The endmembers form the vertices of a simplex, while all possible mixed pixels consisting of the considered endmembers are located between the vertices.

Another group of unmixing methods also covered in more detail in [16] are sparse regression and sparse coding approaches. It is known from compressive sensing theory that real-world signals can be represented sparsely in other transform domains [35]. This fact can be beneficially combined with another fact present in hyperspectral image analysis: there are large spectral libraries available containing hundreds of thousands of pure spectra such as the USGS (United States Geological Survey) spectral library [74, 145]. In a semi-supervised fashion, the endmember matrix $\mathbf{M}$ is assigned the full library, and then it is known that the abundance matrix $\mathbf{A}$ is sparse and consists of many zeros, as only few pure spectra are necessary to reconstruct the acquired pixel spectra. The calculation of $\mathbf{A}$ is conducted by the methods given for instance in [15, 67, 68].

While the methods described so far are only based on spectral information, i.e., the outcome of the procedure is the same regardless if the pixel spectra in the image matrix $\mathbf{Y}$ are permuted or not, spatial information has been exploited in rather few methods. Shi et al. [131] give a review of unmixing methods incorporating spatial information; it must be stated that while many EEAs have been extended by spatial information, the number of abundance estimation algorithms using spatial information is still rather low. By considering Fig. 3.3, however, it is intuitively clear that spatial information has great potential for improving the spectral unmixing abundance estimation. Incorporating the spatial arrangement of the image spectra makes it possible to take the spatial structure into account that is lost when only the isolated pixel spectra would be considered.

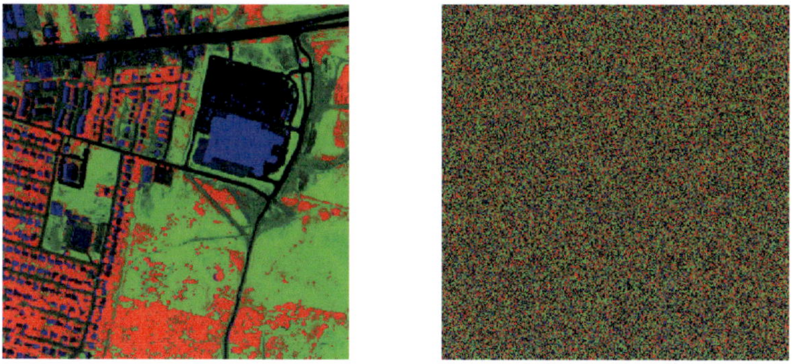

Figure 3.3 Left: Urban image (Sec. 5.1.1) ground truth with four endmembers: grass (green), trees (red), roof (blue), asphalt (black). Right: spatial permutation of the left image [116].

In the following two Sections 3.2.1 and 3.2.2, some endmember extraction algorithms and unsupervised unmixing methods will be briefly explained. These are the existing procedures on which the newly developed methods build on or that will be used for comparison in the results (Chapter 5).

3.2.1 Supervised Unmixing – Endmember Estimation

Explanations of various endmember extraction algorithms can be found, e.g., in [16, 17, 114]. The two EEAs used in this thesis are N-FINDR [150] and simplex identification via split augmented Lagrangian (SISAL) [14].

N-FINDR first applies an orthogonal subspace projection (OSP) along the spectral domain, reducing the spectral dimensionality to $R-1$. Subsequently, all pixel spectra one after the other are put in place of each column of $\mathbf{M}$, and every time, the simplex volume (cf. Fig. 3.2) is calculated. The pixel spectra combination providing the largest volume is considered as the true endmember matrix. When there are pure pixels, this procedure will provide the true endmembers. The algorithm is run with different random initializations, though, to provide an estimate for the case no pure endmembers are present.

SISAL also applies dimensionality reduction first, after which the (reduced) endmember matrix $\mathbf{M}$ is square. This enables its inversion and although it does not yield a convex optimization problem, it leads to suboptimal solutions of good quality. SISAL is based on finding a simplex of minimum volume.

3.2.2 Unsupervised Unmixing

Although designed primarily for endmember estimation, minimum volume constrained NMF (MVC-NMF) [95] is listed in this section. The reason is that the optimization of the NMF objective function does not only provide an endmember estimate, but also an abundance estimate. As known from its name, MVC-NMF aims at finding the minimum volume simplex containing all the pixel spectra and is based on the regularizer

$$\mathcal{R}_{\text{spec}} = \mathcal{R}_{\text{MVC}}(\mathbf{M}) = \frac{1}{2(R-1)!}\det^2\left(\begin{bmatrix}\mathbb{1}_R^{\mathrm{T}}\\ \tilde{\mathbf{M}}\end{bmatrix}\right) . \tag{3.3}$$

This regularizer contains the matrix $\tilde{\mathbf{M}} = (\tilde{\mathbf{m}}_1, \ldots, \tilde{\mathbf{m}}_R) \in \mathbb{R}^{(R-1)\times R}$ defined by

$$\tilde{\mathbf{M}} = \mathbf{P}^{\mathrm{T}}(\mathbf{M} - \boldsymbol{\mu}\mathbb{1}_R^{\mathrm{T}}) . \tag{3.4}$$

The matrix $\mathbf{P} \in \mathbb{R}^{L\times(R-1)}$ is calculated using the $R-1$ most significant principal components of $\mathbf{Y}$. The vector $\boldsymbol{\mu}$ is the data mean which is also calculated from $\mathbf{Y}$. The vector $\mathbb{1}_R$ is an all-ones column vector of length R.

Minimum dispersion constrained NMF (MiniDisCo-NMF), by contrast to MVC-NMF, does not include the simplex concept, but minimizes the intra-spectra variances of all endmembers. The regularizer is calculated by

$$\begin{aligned}\mathcal{R}_{\text{spec}} &= \mathcal{R}_{\text{MiniDisCo}}(\mathbf{M}) \\ &= \operatorname{tr}\left(\left(\mathbf{M} - \frac{1}{L}\mathbb{1}_{LL}\mathbf{M}\right)^{\mathrm{T}}\left(\mathbf{M} - \frac{1}{\mathbf{M}}\mathbb{1}_{LL}\mathbf{M}\right)\right) \\ &= \operatorname{tr}(\mathbf{M}^{\mathrm{T}}\mathbf{M}) - \frac{1}{L}\operatorname{tr}(\mathbf{M}^{\mathrm{T}}\mathbb{1}_{LL}\mathbf{M}) \end{aligned} \tag{3.5}$$

with $\mathbb{1}_{LL}$ being a matrix of ones of size $L \times L$. The goal of the regularizer (3.5) is to promote flat endmember spectra, thus guiding the minimization procedure to the correct solution. Be reminded that data fidelity objective function (3.1) is bi-convex; therefore, local optima exist in both MVC-NMF and MiniDisCo-NMF.

The minimum volume enclosing simplex (MVES) unmixing method [24] expresses the simplex volume in a low-dimensional space which similarly to SISAL also leads to an invertible (low-dimensional) endmember matrix. One major difference between SISAL and MVES is that SISAL requires the dimensionality reduction to take place before applying the algorithm, while for MVES, it is inherently contained in the method. Using an inverse matrix in MVES enables the conversion into a sequence of linear optimization problems. Although the total objective function is not convex, MVES is guaranteed to find the true endmembers in the case pure pixels of each endmember are present in the image.

As the following unmixing method is especially interesting for the methods that have been developed newly in the course of the present thesis, it will be explained in greater detail.

Zymnis et al. [163] introduce a total variation (TV)-like spatial regularization into spectral unmixing. The well-known TV method has been presented by Rudin, Osher and Fatemi [127] mainly as an image denoising method and numerous extensions exist nowadays.

Linear filtering for denoising purposes removes mainly high-frequency coefficients, therefore eliminating (or at least lowering) the noise. This, however, also smoothes sharp edges, as all frequencies are required to represent signal jumps. For this reason, TV follows a different approach: the goal is to approximate the measured (noisy) signal as well as possible, while on the other hand introducing the constraint that differences of neighboring values of the denoised signal should have as small values as possible. The latter constraint can be reformulated by requiring the spatial gradient to be as small as possible. For 2D signals, this results in the following minimization objective function (see, e.g., [124]):

$$Q_{\mathrm{TV}} = \frac{1}{2}\sum_{x,y}\left(u_{xy} - z_{xy}\right)^2 + \mu \sum_{x,y}\sqrt{\left|u_{x+1,y} - u_{xy}\right|^2 + \left|u_{x,y+1} - u_{xy}\right|^2}\,. \tag{3.6}$$

Here, u_{xy} is the pixel at position (x, y) of the uncorrupted signal $\mathbf{U}$ that will be approximated, while $\mathbf{Z}$ denotes the measured signal. The regularization parameter μ balances both summands. The total variation norm $\sum_{x,y}\sqrt{\left|u_{x+1,y} - u_{xy}\right|^2 + \left|u_{x,y+1} - u_{xy}\right|^2}$ enforces a constant signal sequence by minimizing the absolute gradient. All image elements u_{xy} can be arranged to a vector of size $X \cdot Y \times 1$. The square root term is the discretization of the absolute gradient, and the square root terms at all pixel positions can also be arranged to a vector of respective size [124]. The ℓ_1 norm of this vector is just the sum of all its elements, as represented in (3.6).

Applying TV to spectral unmixing, Zymnis et al. [163] propose the objective function

$$Q = \|\mathbf{MA} - \mathbf{Y}\|_{\mathrm{F}}^2 + \mu \sum_{i=1}^{I} \sum_{j \in \mathcal{N}(i)} \|\mathbf{a}_i - \mathbf{a}_j\|_1\,. \tag{3.7}$$

Doing so requires the abundance maps $\mathbf{A}_r$ of all endmembers to have as low spatial gradient values as possible. Note that the matrices $\mathbf{A}_r \in \mathbb{R}^{X \times Y}$ hold the abundances of endmember r at each pixel, rearranged to the original image dimensions. $\mathbf{a}_i$ stands for one column of the (unfolded) abundance matrix $\mathbf{A} \in \mathbb{R}^{R \times I}$, i.e., the abundances of pixel i, wherever in the image it might be located. The pixel neighborhood $\mathcal{N}(i)$ denotes the set of indices of the incorporated neighbors of pixel i. This regularization is beneficial, as natural images are likely to have constant image regions and jumps. One downside of the 2D (3.6) objective function, however, is the fact that linearly increasing/decreasing image values will be approximated by steps instead of a smooth shape. For circumventing this, the TV method will be extended by methods incorporating different spatial derivatives of the abundance maps and different norms in Sec. 4.1. These enable piecewise linear transitions.

3.3 Hyperspectral Image Denoising

This section gives a brief introduction of hyperspectral image denoising in preparation of the newly developed denoising algorithm that will be explained in Sec. 4.7.

The easiest way to denoise a multidimensional image is to split up the image into several one- or two-dimensional subparts and denoise them with common denoising techniques. In the case of three-dimensional hyperspectral images having two spatial and one spectral dimension, the pixel spectra can be regarded as one-dimensional signals and treated as such. On the other hand, the two-dimensional images at each wavelength can be treated separately with conventional or sophisticated image denoising methods such as total variation (TV [127]), non-local means [21], block-matching and 3D filtering (BM3D, [34]) or their variants. Such approaches, however, do not take into account the significant amount of redundancy contained in the multidimensionality.

For 3D hyperspectral images, many methods have been developed that take advantage of the full dimensionality of the data. One approach is to extend the conventional 2D TV model [127], which is known for its edge-preserving property, to three dimensions by adding a third spectral dimension that is filtered with 1D TV [158]. This approach neglects spatial and spectral variations of noise and signal; therefore, Yuan et al. [155] proposed a spectral–spatial adaptive hyperspectral TV (SSAHTV) denoising algorithm taking these into account. Qian et al. [121] follow a different denoising approach by extending the non-local means filter [21] to 3D images. The anisotropic diffusion method by Perona and Malik [110] has also been extended to hyperspectral images by Wang et al. [148].

An approach different to the aforementioned methods is followed by Lam et al. [76]. They perform principal component analysis (PCA) on the image to reduce the spectral dimensionality and capture only the main variance, disregarding the variance caused by noise. The coefficients associated with the principal components form 2D images; one for each principal component. These coefficient images are denoised independently with common 2D denoising techniques, and PCA backprojection yields the denoised hyperspectral image. A procedure that can be considered complementary to this approach is described in [30]. It is based on low-noise images and aims at removing the remaining noise. For preserving the physical structure contained in the image, only the principal components describing small proportions of the image variance are denoised, i.e., the ones that are completely disregarded by the method in [76]. The denoising is done by wavelet shrinkage.

The hybrid spatial-spectral noise reduction (HSSNR) scheme described in [106] also deals with low-noise images. First, the spectral derivative is calculated which results in an increased noise level. The resulting signal is denoised independently in the spatial and spectral domains and integrated spectrally afterwards. This provides the denoised datacube.

Following the natural approach of treating the hyperspectral datacube as a third-order tensor also leads to denoising algorithms. Letexier and Bourennane [78] have developed such a technique that is based on the multidimensional Tucker3 decomposition and multidimensional Wiener filtering. This technique is compared with other tensor methods [84]. Xu et al. [151] model the denoising problem as a Bayesian least squares optimization problem and propose a Monte Carlo sampling technique to determine the a posteriori distribution. The recently proposed noise adjusted iterative low-rank matrix approximation (NAILRMA) method [58] is based on the observation that hyperspectral images commonly are low rank and therefore the goal is to find a low-rank approximation of the noisy image.

4 Novel Methods

This chapter contains the descriptions of the methods newly developed during the work on this thesis. Starting from unmixing methods enforcing local spatial abundance smoothness in the following section, Sec. 4.2 presents a method taking larger pixel regions into account. This leads to a calculation time saving and in some cases, especially in the presence of noise, improves the unmixing results. Section 4.3 presents the use of image pyramids for accelerating existing unmixing methods. Its mere goal is the acceleration of the unmixing. These two unmixing methods innovatively consider image matrix reduction along the spatial domain, and not, as commonly done, along the spectral domain. The newly developed unmixing method described in Sec. 4.4 incorporates the full adjacency of both spatial and spectral image elements by applying the 3D discrete cosine transform (3D DCT) before unmixing. This method even incorporates the full adjacency in the data cube. As an example of other orthogonal transforms sharing the beneficial properties of the 3D DCT and that can be used for using the 3D information is the singular value decomposition (SVD). The 3D unmixing method based on SVD is described in Sec. 4.5.

The chapter is closed by a newly developed denoising algorithm. While it is easy to apply and runs much faster than state-of-the-art algorithms, it offers similar, in many cases even better, denoising capability.

4.1 Unmixing Methods Incorporating Spatial Abundance Derivatives

Starting from the total variation (TV) method that has been introduced briefly in Section 3.2.2, newly developed unmixing methods that promote spatial smoothness will be presented. Most NMF unmixing algorithms alternatingly minimize the objective function, for instance (3.2), for abundance matrix $\mathbf{A}$ and endmember matrix $\mathbf{M}$. For this reason, a convention used throughout the thesis is introduced now. For the matrix $\mathbf{A}$, the number of iterations is P, as it is possible to run multiple iterations for one matrix before updating the other. For $\mathbf{M}$, the number of iterations is M. The total number of $\mathbf{A}/\mathbf{M}$ updates is denoted by T. This means that P abundance iterations and M endmember iterations are performed, and this P/M sequence is repeated T times in total.

The following detailed TV description is based on the original paper [163]. Considering only one pixel, the optimization problem deduced from (3.7) is given by

$$\begin{aligned} &\text{minimize } \|\mathbf{y}_i - \mathbf{M}\mathbf{a}_i\|_{\mathrm{F}}^2 + \mu \sum_{j \in \mathcal{N}(i)} \|\mathbf{a}_i - \mathbf{a}_j\|_1 \\ &\text{subject to } a_{r,i} \geq 0, \ \sum_r a_{r,i} = 1 \,. \end{aligned} \tag{4.1}$$

It must be noted that this problem contains the vector ℓ_1 norm of the difference $(\mathbf{a}_i - \mathbf{a}_j)$, i.e., the sum of absolute vector elements. As the absolute value is not differentiable, a different method has to be used for applying a descent method for minimizing (4.1). The subgradient method described for example in [132] is an optimization method capable of dealing with non-differentiable objective functions.

Optimizing for $\mathbf{a}_i$, the other vectors $\mathbf{a}_j$ remain constant during this procedure. The vector

$$\mathbf{g}_i = 2\mathbf{M}^{\mathrm{T}}\mathbf{M}\mathbf{a}_i - 2\mathbf{M}^{\mathrm{T}}\mathbf{y}_i + \mu\mathbf{s}_i \tag{4.2}$$

is a subgradient of the minimization problem (4.1). It mainly results from differentiation of its differentiable part, while the vector $\mathbf{s}_i \in \mathbb{R}^R$ is calculated by $\mathbf{s}_i = \sum_{j \in \mathcal{N}(i)} \operatorname{sgn}(\mathbf{a}_i - \mathbf{a}_j)$, where sgn denotes the sign function. Similarly to a gradient descent method, the objective function value is iteratively minimized using

$$\mathbf{a}_i^{(p+1)} := \left(\mathbf{a}_i^{(p)} - \delta^{(p)}\mathbf{g}_i\right)_{\mathrm{P}} \,. \tag{4.3}$$

The variable p stands for the iteration number, while $\delta^{(p)}$ is the current step size. $(\cdot)_{\mathrm{P}}$ denotes projection on the probability simplex for ensuring compliance of the solution with the STO and the NN constraints (Sec. 2.1.1). In contrast to [163], not the projection algorithm provided therein, but the projection algorithm given in [30] is used for all methods described in this chapter. It is extended to a matrix version enabling the simultaneous processing of the whole abundance matrix, therefore considering all pixels at the same time and computing very fast.

All columns of $\mathbf{A}$ can be updated simultaneously. Zymnis et al. [163] state that this is due to the principle of primal decomposition. Alternatively, it can be explained by the gradient calculation of scalar functions that depend on a matrix (Sec. 2.2.4): the function is differentiated by each matrix element, while all other elements are considered constants. The results of all element derivatives are put into the same order as the respective elements in the original matrix. Concatenating all vectors $\mathbf{s}_i$ to the matrix $\mathbf{S}$ leads to the matrix update

$$\mathbf{A}^{(p+1)} := \left(\mathbf{A}^{(p)} - \delta^{(p)}(2\mathbf{M}^{\mathrm{T}}\mathbf{M}\mathbf{A}^{(p)} - 2\mathbf{M}^{\mathrm{T}}\mathbf{Y} + \mu\mathbf{S}^{(p)})\right)_{\mathrm{P}} . \tag{4.4}$$

After updating $\mathbf{A}$ according to this rule several times for constant $\mathbf{M}$, the matrix $\mathbf{M}$ is updated by solving the problem

$$\text{minimize } \|\mathbf{Y} - \mathbf{M}\mathbf{A}\|_{\mathrm{F}}^2 \tag{4.5}$$
$$\text{subject to } \mathbf{M} \geq \mathbf{0} . \tag{4.6}$$

This time, the gradient descent steps

$$\mathbf{M}^{(m+1)} := \left(\mathbf{M}^{(m)} - \delta^{(m)}(\mathbf{M}^{(m)}\mathbf{A} - \mathbf{Y})\mathbf{A}^{\mathrm{T}}\right)_{+} \tag{4.7}$$

can be applied. The operator $(\cdot)_+$ projects negative matrix values to zero (NN constraint, Sec. 2.1.1), while the step size is defined to

$$\delta^{(m)} = \frac{1}{\|\mathbf{A}\mathbf{A}^{\mathrm{T}}\|_2} , \tag{4.8}$$

where $\|\cdot\|_2$ denotes the matrix spectral norm.

As the elements of $\mathbf{Y}$ can have values of different magnitudes depending on the input image, it is reasonable to use the nonsummable diminishing step size rule [11] for the update of $\mathbf{A}$. The method is guaranteed to converge when $\delta^{(p)}$ is chosen to $\delta^{(p)} = \frac{k}{p}$ [117]. The constant k is chosen feasibly; however, there are more effective ways of adjusting the step size at each iteration which are discussed in Sec. 4.6.1.

The TV method applied to 1D or 2D signals is known to suffer from the so-called staircase effect: While signal jumps are preserved well, linear or curvy image regions are approximated by steps. This effect is illustrated in Fig. 4.1 for a 1D signal. A large number of approaches for mitigating or even eliminating the staircase effect have been proposed, see, e.g., [20, 62]. One option is to use not the first, but higher order spatial derivatives. This can be combined with penalizing the values of the respective derivative not by the vector ℓ_1 norm as in (3.6), but for example by the ℓ_2 norm [124]. Apart from the TV denoising result, Fig. 4.1 also illustrates the denoising using the first derivative and the ℓ_2 norm. Additionally, it shows the use of combinations of the second spatial derivative and the ℓ_1/ℓ_2 norms. It becomes obvious that in the case of the first spatial derivative, the ℓ_2 norm penalizes high values of the derivative harder than the ℓ_1 norm, resulting in a smoothing of the edges that TV preserves so well. On the other hand, there is no staircase effect in linear signal regions. The second spatial derivative combined with the ℓ_1 norm results in a piecewise linear denoised signal, while the application of the ℓ_2 norm to the second derivative tends to introduce small oscillations. Figure 4.1 raises the questions how regularizers based on the TV-like objective (3.7), but combining the first/second derivative and the ℓ_1/ℓ_2 norm behave when transferred to spectral unmixing. For this reason, the respective objective functions will be presented in the following sections.

As a first step, a regularizer replacing the TV-like one of (3.7) will be explained that incorporates the ℓ_2 norm of the second spatial derivative. Its goal is to approximate linear changes in the abundance maps as closely as possible, without introducing artifacts such as the staircase effect. The method has been developed by the author and was published in [176].

For an arbitrary continuous 2D function $f(x, y)$, the second derivative is its Hessian

$$H(f) = \begin{bmatrix} \frac{\partial^2 f}{\partial x^2} & \frac{\partial^2 f}{\partial x \partial y} \\ \frac{\partial^2 f}{\partial y \partial x} & \frac{\partial^2 f}{\partial y^2} \end{bmatrix} . \tag{4.9}$$

Transferring this to spatial abundance regularization, the reader be reminded that the abundance map $\mathbf{A}_r$ is a matrix of size $X \times Y$ containing the

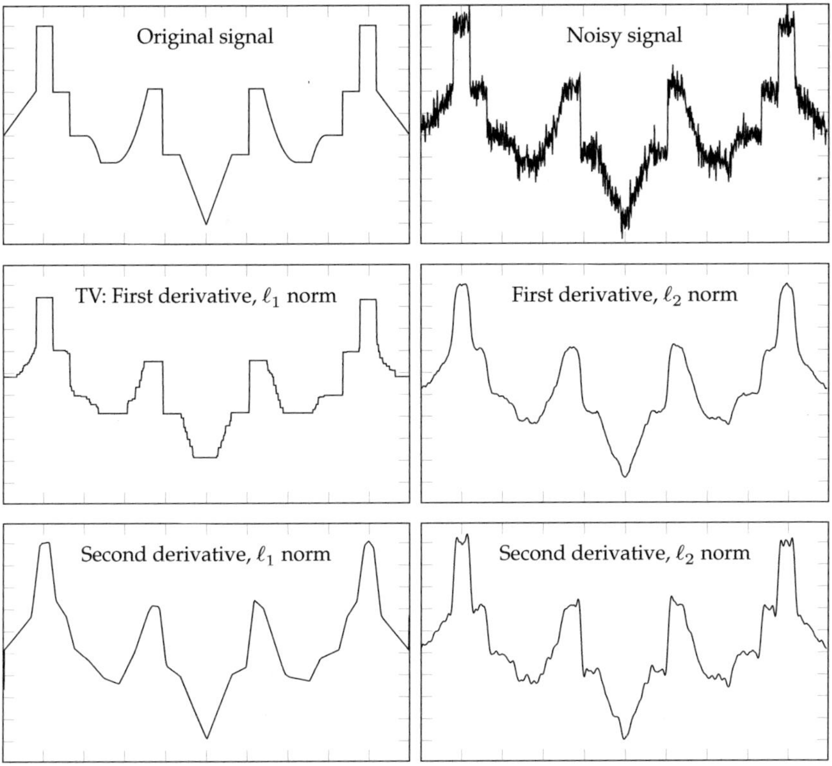

Figure 4.1 Denoising of a noisy signal of length 1000 with combinations of the first and second spatial derivatives combined with their ℓ_1/ℓ_2 norm.

pixelwise abundances of substance r in its respective spatial arrangement. Calculating the Frobenius norm of the Hessian of $\mathbf{A}_r$ at pixel position (x, y) results to

$$\|H(\mathbf{A}_r)(x,y)\|_\mathrm{F} = \sqrt{\mathbf{A}^2_{r,xx}(x,y) + \mathbf{A}^2_{r,xy}(x,y) + \mathbf{A}^2_{r,yx}(x,y) + \mathbf{A}^2_{r,yy}(x,y)}\,. \tag{4.10}$$

The term $\mathbf{A}_{r,xx}$ denotes the second derivative of $\mathbf{A}_r$ along direction x. The derivatives along the respective other directions are denoted by the terms $\mathbf{A}_{r,xy}$, $\mathbf{A}_{r,yx}$ and $\mathbf{A}_{r,yy}$. In the following, the second derivative regularizer based on its Frobenius norm will be abbreviated as Hessian (HS) regularization.

Adding the terms of (4.10) for all endmembers r and all pixels provides the second-order regularizer

$$\mathcal{R}_{\mathrm{HS}}\left(\mathbf{A}\right) = \sum_{r=1}^{R}\sum_{x=1}^{X}\sum_{y=1}^{Y} \|H(\mathbf{A}_r)(x,y)\|_{\mathrm{F}} \,. \tag{4.11}$$

The final HS objective function results to

$$Q_{\mathrm{HS}} = \|\mathbf{MA} - \mathbf{Y}\|_{\mathrm{F}}^2 + \mu_{\mathrm{spat}}\mathcal{R}_{\mathrm{HS}}\left(\mathbf{A}\right) \,. \tag{4.12}$$

When optimizing for pixel $\mathbf{a}_i$, the pixels in its neighborhood which are required for calculating the spatial derivatives $\mathbf{A}_{r,xx}$, $\mathbf{A}_{r,xy}$ etc. remain constant. The objective function (4.12) inherits the bi-convexity of the pure data fidelity term $\|\mathbf{MA} - \mathbf{Y}\|_{\mathrm{F}}^2$ (3.1). The derivatives $\mathbf{A}_{r,xx}$ etc. are basically affine combinations of the abundance values (i.e., the elements of $\mathbf{A}$), as they are only added and subtracted, which will be shown later in this section. The Frobenius norm is a matrix norm and therefore convex, as all norms are convex [19, Sec. 3.1.5]. According to [19, Sec. 3.2.2], a composition of an affine mapping and a convex function is also convex. Therefore, (4.12) is convex in $\mathbf{A}$. In contrast to the TV-like regularizer (3.7), the objective function (4.12) is differentiable. For this reason, the gradient projection method [72] is used.

The full HS optimization problem is given by the objective function (4.12) and the NN and STO constraints:

$$\begin{aligned} &\text{minimize } \|\mathbf{MA} - \mathbf{Y}\|_{\mathrm{F}}^2 + \mu_{\mathrm{spat}}\mathcal{R}_{\mathrm{HS}}\left(\mathbf{A}\right) \\ &\text{subject to } \mathbf{A} \geq 0,\ \sum_{\forall r} a_{r,i} = 1 \,, \end{aligned} \tag{4.13}$$

where the inequality $\mathbf{A} \geq 0$ means that each element of $\mathbf{A}$ has to be ≥ 0. In [176], the minimization is solved with square summable but not summable step sizes fulfilling

$$\sum_{p=1}^{\infty} \delta^{(p)} = \infty,\ \sum_{p=1}^{\infty} \left(\delta^{(p)}\right)^2 < \infty \,. \tag{4.14}$$

The step size is chosen to be the same as the one of the TV-like optimization:

$$\delta^{(p)} = \frac{k}{p} \,. \tag{4.15}$$

When updating the abundance values according to

$$\mathbf{A}^{(p+1)} := \left(\mathbf{A}^{(p)} - \delta^{(p)} \frac{\partial Q_{\mathrm{HS}}}{\partial \mathbf{A}} \right)_{\mathrm{P}} , \tag{4.16}$$

the gradient $\frac{\partial Q_{\mathrm{HS}}}{\partial \mathbf{A}}$ of Q_{HS} with respect to $\mathbf{A}$ has to be calculated. While doing so, care should be taken. According to [90], the Frobenius norm of the Hessian (4.11) is rotationally invariant. This, however, only holds in the continuous case. When approximating the continuous second spatial derivatives by the forward differences

$$\begin{aligned}
\mathbf{A}_{r,xx}(x,y) &= \mathbf{A}_r(x,y) - 2\mathbf{A}_r(x+1,y) + \mathbf{A}_r(x+2,y) \\
\mathbf{A}_{r,xy}(x,y) &= \mathbf{A}_r(x,y) - \mathbf{A}_r(x+1,y) - \mathbf{A}_r(x,y+1) + \mathbf{A}_r(x+1,y+1) \\
\mathbf{A}_{r,yx}(x,y) &= \mathbf{A}_{r,xy}(x,y) \\
\mathbf{A}_{r,yy}(x,y) &= \mathbf{A}_r(x,y) - 2\mathbf{A}_r(x,y+1) + \mathbf{A}_r(x,y+2) \,,
\end{aligned} \tag{4.17}$$

wavelike artifacts are introduced into the abundance estimation, see Fig. 4.2. For this reason, the central differences [8]

$$\begin{aligned}
\mathbf{A}_{r,xx}(x,y) &= \mathbf{A}_r(x-1,y) - 2\mathbf{A}_r(x,y) + \mathbf{A}_r(x+1,y) \\
\mathbf{A}_{r,xy}(x,y) &= \mathbf{A}_r(x,y+1) - \mathbf{A}_r(x,y) - \mathbf{A}_r(x-1,y+1) + \mathbf{A}_r(x-1,y) \\
\mathbf{A}_{r,yx}(x,y) &= \mathbf{A}_r(x+1,y) - \mathbf{A}_r(x,y) - \mathbf{A}_r(x+1,y-1) + \mathbf{A}_r(x,y-1) \\
\mathbf{A}_{r,yy}(x,y) &= \mathbf{A}_r(x,y-1) - 2\mathbf{A}_r(x,y) + \mathbf{A}_r(x,y+1)
\end{aligned} \tag{4.18}$$

will be used for HS unmixing. Their usage does not introduce any artifacts. Having defined the discrete approximation of the continuous derivatives, it is now possible to calculate the regularizer's (4.11) gradient $\frac{\partial \mathcal{R}_{\mathrm{HS}}}{\partial \mathbf{A}}$. This is a matrix of the same size as $\mathbf{A}$. Considering the abundance of the r-th

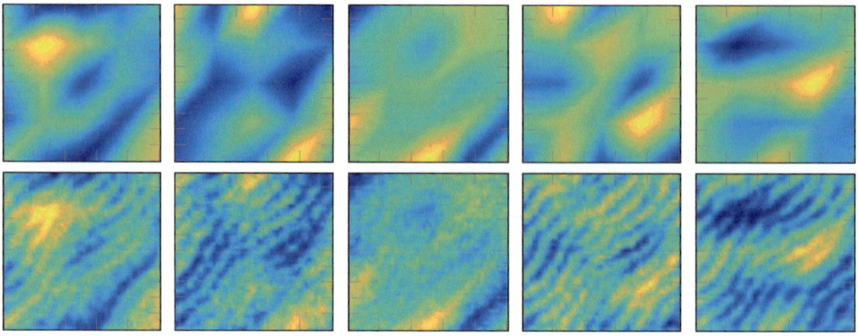

Figure 4.2 Qualitative visualization of the wavelike artifacts caused by the use of the forward differences (4.17) for HS regularization. Shown are the five abundance maps of an artificial image of size 50×50. First line: ground truth, second line: abundance estimation using (4.12).

endmember at position (x, y), one element of the gradient is given by

$$\left(\frac{\partial \mathcal{R}_{\text{HS}}}{\partial \mathbf{A}}\right)_r (x, y) = \tag{4.19}$$
$$\Big\{ 10\mathbf{A}_r(x, y) - 3\mathbf{A}_r(x, y+1) - 3\mathbf{A}_r(x, y-1) - 3\mathbf{A}_r(x-1, y)$$
$$-3\mathbf{A}_r(x+1, y) + \mathbf{A}_r(x-1, y+1) + \mathbf{A}_r(x+1, y-1) \Big\} \,/$$
$$\Big\{ \left[\mathbf{A}_r(x-1, y) - 2\mathbf{A}_r(x, y) + \mathbf{A}_r(x+1, y)\right]^2$$
$$+ \left[\mathbf{A}_r(x, y+1) - \mathbf{A}_r(x, y) - \mathbf{A}_r(x-1, y+1) + \mathbf{A}_r(x-1, y)\right]^2$$
$$+ \left[\mathbf{A}_r(x+1, y) - \mathbf{A}_r(x, y) - \mathbf{A}_r(x+1, y-1) + \mathbf{A}_r(x, y-1)\right]^2$$
$$+ \left[\mathbf{A}_r(x, y-1) - 2\mathbf{A}_r(x, y) + \mathbf{A}_r(x, y+1)\right]^2 \Big\}^{(1/2)} .$$

The denominator can become zero when the abundances of endmember r at the respective pixel and around it have the same value. Although this is numerically unlikely to happen, the gradient is set to zero in such a case, because the second spatial derivative is already zero and the respective abundance value should remain unchanged, see the update (4.16).

Having extended the TV regularization to ℓ_2 penalization of the second derivative, the two other combinations of first/second derivative and

ℓ_1/ℓ_2 norm are still open. They have been developed newly as well and are published in [166].

The regularizer of the first spatial derivative combined with the ℓ_2 norm is given by

$$\mathcal{R}_{\mathrm{D1-L2}} = \sum_{i=1}^{I} \sum_{j \in \mathcal{N}(i)} \left\| \mathbf{a}_i - \mathbf{a}_j \right\|_2^2 , \tag{4.20}$$

where the square has been added to penalize large regularizer values harder. It also leads to easier calculation of the gradient. The regularizer (4.20) can be rewritten

$$\mathcal{R}_{\mathrm{D1-L2}} = \sum_{i=1}^{I} \sum_{j \in \mathbb{N}(i)} \sum_{r=1}^{R} (a_{ri} - a_{rj})^2 , \tag{4.21}$$

which leads to the gradient

$$\left(\frac{\partial \mathcal{R}_{\mathrm{D1-L2}}}{\partial \mathbf{A}} \right)_{ri} = \sum_{j \in \mathbb{N}(i)} 2 \left(a_{ri} - a_{rj} \right) \tag{4.22}$$

at matrix position ri.

Application of the ℓ_1 norm to the second spatial derivative yields

$$\mathcal{R}_{\mathrm{D2-L1}}(\mathbf{A}) = \sum_{r=1}^{R} \sum_{x=1}^{X} \sum_{y=1}^{Y} \left\| H\left(\mathbf{A}_r(x,y)\right) \right\|_1 . \tag{4.23}$$

Like the TV regularizer (4.1), this regularizer also is not differentiable. The subgradient method provides the matrix $\mathbf{S}_{\text{D2-L1}}$ with elements

$$\begin{aligned}
\left(\mathbf{S}_{\text{D2-L1}}\right)_r (x,y) &= -2\,\mathrm{sgn}\left(\mathbf{A}_r\left(x-1,y\right) - 2\mathbf{A}_r\left(x,y\right) + \mathbf{A}_r\left(x+1,y\right)\right) \\
&+ \mathrm{sgn}\left(\mathbf{A}_r\left(x,y+1\right) - \mathbf{A}_r\left(x,y\right) - \mathbf{A}_r\left(x-1,y+1\right) + \mathbf{A}_r\left(x-1,y\right)\right) \\
&+ \mathrm{sgn}\left(\mathbf{A}_r\left(x+1,y\right) - \mathbf{A}_r\left(x,y\right) - \mathbf{A}_r\left(x+1,y-1\right) + \mathbf{A}_r\left(x,y-1\right)\right) \\
&- 2\,\mathrm{sgn}\left(\mathbf{A}_r\left(x,y-1\right) - 2\mathbf{A}_r\left(x,y\right) + \mathbf{A}_r\left(x,y+1\right)\right) .
\end{aligned} \tag{4.24}$$

The HS results are briefly analyzed in Sec. 4.6. For more analyses, see [176]. The results of the other regularizers are given in [166].

4.2 Unmixing of Clustered Image (UNCLI)

In Section 4.1, unmixing methods exploiting the characteristics of spatial abundance derivatives have been presented. These are based on the local neighborhood around a considered pixel such as the 8-connected pixels, see Fig. 4.3.

(a) TV 8-connected neighborhood **(b)** HS neighborhood

Figure 4.3 Pixel neighborhoods used for spatial regularization. The central pixel is shown in red, the other incorporated pixels in blue. The HS neighborhood results from the central differences (4.18).

When analyzing real hyperspectral images in more detail, it becomes obvious that spatial information within even larger image regions can be used beneficially. Figure 4.4 shows the ground truth of the Urban image (see Sec. 5.1.1) which illustrates a remote view of an urban scene. The image mainly contains roof, asphalt, grass and trees. It can be seen that many pixels have very similar abundances, e.g., the asphalt pixels. The figure also shows the spectra of all pixels that consist of at least 97 % asphalt. When considering the fact that hyperspectral image processing is time-consuming, it becomes clear that it would be beneficial to reduce the size of the image matrix $\mathbf{Y}$ by some operation, such that less pixels (i.e., columns of the image matrix) have to be considered without losing much accuracy. Such an approach is a technique of *object-space* reduction [42], as it reduces the number of objects, i.e., pixels. By contrast, methods such as PCA along the spectral domain, orthogonal subspace projection (OSP) [56] or maximum noise fraction (MNF) [52] aim at reducing the *feature-space*, i.e., the number of bands, and not the number of pixels.

One natural approach of reducing the image matrix size is to cluster the image pixels according to their similarity in the sense of some measure.

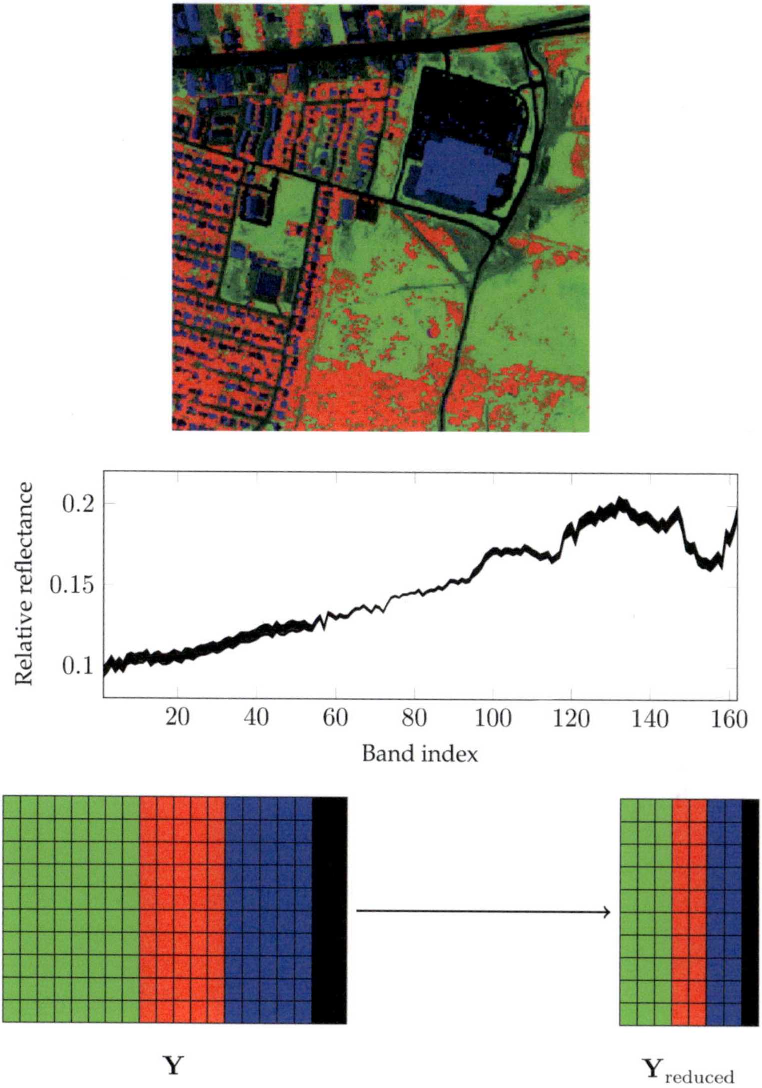

Figure 4.4 Top: Urban image (Sec. 5.1.1) ground truth with four endmembers: grass (green), trees (red), roof (blue), asphalt (black). Middle: spectra of pixels with ≥97 % asphalt abundance. Bottom: image matrix $\mathbf{Y}$ and reduced image matrix $\mathbf{Y}_{\text{reduced}}$.

This is not only useful because most images have smaller or larger regions containing very similar abundances, but also when spectral regularization is incorporated, e.g., MiniDisCo-NMF [63], MVC-NMF [95] (see Sec. 3.2.2), minimum endmember-wise distance constrained NMF (MewDC-NMF) [94] or minimum distance constrained NMF (MDC-NMF) [154]. In general, updating the abundance matrix $\mathbf{A}$ in the pure data fidelity objective function (3.1) is much more demanding than updating $\mathbf{M}$ because of the matrix sizes (e.g., 10×10000 vs. 200×10). When using spectral regularization for endmember extraction, the abundance matrix is more or less a by-product that is not of significant interest. In this case, it is of great benefit to use a compressed abundance matrix to get the desired endmember matrix $\mathbf{M}$ more quickly. As will be shown in the results (Sec. 5.4), the endmember estimation results additionally are significantly improved in the presence of noise. Even when not performing endmember estimation, but unsupervised unmixing, where the calculated abundance matrix is of interest, its estimate can be very precise when suitable clustering methods are used. This means that there is a large gain in calculation time, while the abundance and endmember estimation results are equally as good or even better. The new approach will be called unmixing of clustered image (UNCLI). As it can be combined with the vast majority of unmixing algorithms, it is a meta-approach. Its principle was published by Bauer et al. [175]. It must be stated that the author is not aware of any other object-space reduction methods in the area of spectral unmixing abundance estimation. The brief report [109] describes a similar idea, but does not provide a thorough investigation of the results.

Although clustering algorithms have been used previously in quite a few hyperspectral image processing steps for purposes such as compression [98] and classification [140], they have not yet been used for the acceleration of the unmixing process by reducing the amount of data. Regarding the existing literature in the area of spectral unmixing, the paper of Lu et al. [88] is especially interesting. They use the structural information present in an image for improving the unmixing results with their structure constrained sparse NMF (CSNMF) method. The k-means algorithm is used to divide the image into a number of clusters and for each pixel, a regularization term ensuring that the pixel's abundances are close to the cluster centroid's abundances is included. Additionally, a sparsity regularizer is incorporated. CSNMF outperforms state-of-the-art algorithms like SISAL [14], $L_{1/2}$-NMF [120] and MVES [24] in the case of high noise levels. The calculation of CSNMF, however, is more time consuming than the calculation of the aforementioned algorithms.

4.2.1 UNCLI Procedure

The UNCLI clustering is applied to the original image. It is possible to perform an optional dimensionality reduction step along the wavelength axis before the clustering for even further acceleration. Methods like PCA or MNF can be used for this dimensionality reduction step. Subsequently, the chosen unmixing algorithm is applied to the clustered image. It must be noted that the spectral dimensionality reduction is applied only for accelerating the clustering; unmixing is performed using the full spectral dimensionality. Using the reduced spectral dimensionality for UNCLI could be subject to future work.

When dimensionality reduction is applied, the original image $\mathbf{Y} \in \mathbb{R}^{L \times I}$ is reduced to $\mathbf{Y}' \in \mathbb{R}^{L' \times I}$, $L' < L$. The clustering result is encoded in the segmentation map $\mathbf{C}_{\mathbf{Y}} \in \mathbb{R}^{X \times Y}$. Its elements $\mathbf{C}_{\mathbf{Y},xy} \in \{1, \ldots, C\}$ define each pixel's mapping to one of the C clusters, $C \in \mathbb{N}$. Using this map, the clustered image matrix $\mathbf{Y}_{\text{clust}} \in \mathbb{R}^{L \times C}$ is created. Its columns contain the cluster mean spectra. Although throughout this thesis, no weighting is applied for calculating the mean spectra, it would be possible to use weighted averaging such as the one given in [162].

The following unmixing step yields the calculated matrices $\mathbf{A}_{\text{clust}}$ and $\mathbf{M}$. The segmentation map $\mathbf{C}_{\mathbf{Y}}$ is required for assigning each original pixel contained in the respective cluster the abundance calculated for the cluster mean spectrum. The full UNCLI procedure is given in Algorithm 1.

4.2.2 Used Clustering Methods

For clustering hyperspectral images, several methods have been proposed in the literature. For a brief overview, see, e.g., [107]. In principle, each of the methods described there can be used for the newly developed UNCLI method, as its main goal is to merge data but still achieve an unmixing result quality which is close to the one of full image unmixing.

Algorithm 1 UNCLI procedure.

1) Given $\mathbf{Y}$
2) **if** dimension_reduction==true **then**
   $\mathbf{Y}' = \text{PCA}(\mathbf{Y})$
   Calculate segmentation map $\mathbf{C_Y}$ by clustering $\mathbf{Y}'$
   else
   Calculate segmentation map $\mathbf{C_Y}$ by clustering $\mathbf{Y}$
   endif
3) Calculate cluster matrix $\mathbf{Y}_{\text{clust}}$ using $\mathbf{C_Y}$
4) Perform unmixing with arbitrary method using $\mathbf{Y}_{\text{clust}}$. This yields endmember matrix $\mathbf{M}$ and abundance matrix $\mathbf{A}_{\text{clust}}$
5) Calculate abundance matrix $\mathbf{A}$ of original image from $\mathbf{A}_{\text{clust}}$ using $\mathbf{C_Y}$

It is often not obvious from the clustering procedure which impact the used clustering technique will have on the unmixing result. As long as the number of clusters is not too small (low spatial resolution) or too high (no benefit from clustering), it can be chosen from a relatively wide range, as there is no ideal cluster number which is defined for example by the number of objects present in the scene. The main and only goal is the acceleration of the unmixing process, while providing improved estimations, or at least acceptable unmixing error values. It is also not important if the method clusters adjacent or distant pixels, as long as they have similar spectra. Several clustering methods were tested. They will be described shortly in the following:

RHSEG The recursive hierarchical segmentation (RHSEG) algorithm [144] is an enhancement of the hierarchical segmentation (HSEG) method [143, 144]. It overcomes HSEG's drawback of high computational load in the case of spectral clustering by using a divide-and-conquer technique. The original HSEG method enables both spatial and spectral clustering which is balanced by the parameter sw. For $sw = 0$, the pixels are clustered exclusively by spatial neighborhood considerations, while $sw = 1$ means pure spectral clustering. In the latter case, the computational load is significant, as all pixel spectra have to be considered instead of just more or less adjacent ones. In general, the dissimilarity criterion contained in the algorithm can be chosen to any distance measure between two spectra.

For the analyses presented in Sec. 5.4, Euclidean distance (ED) and spectral angle (SA, Sec. 2.2.3) will be considered as dissimilarity criteria. For the RHSEG algorithm, the implementation provided in [102] is used. Each band of the hyperspectral image that is to be clustered is stored on the hard disk in a separate file. Also, a parameter file is saved. The compiled RHSEG .exe-file is called from within *MATLAB* with shell commands and reads the provided files. After the calculation, the results are also saved as an image and imported in *MATLAB*.

k-means The well-known k-means algorithm [39] is also used for spectral clustering: Given k initial cluster centers, each data point is assigned to the cluster center it has the least distance to. Iteratively, new cluster centers are calculated, the other data points are assigned again to the new cluster centers and so on. Eventually, the algorithm converges. As implementation, *MATLAB*'s `kmeans` method was used selecting the distance measures squared Euclidean distance and correlation.

ISODATA (iterative self-organizing data analysis technique) [5] is an algorithm based on k-means. It does not need an a priori definition of the number of clusters. Only a rough range of possible numbers has to be defined, the exact number is determined by the algorithm. For ISODATA, ED and SA are used as distance measures. ISODATA is implemented in *MATLAB* based on [70].

Superpixel Another used clustering method is the graph-based image segmentation method by Felzenszwalb and Huttenlocher [47]. It will be called superpixel (SP) in the following, as it belongs to the class of superpixel methods. SP also uses Euclidean distance and spectral angle. The implementation is provided in [41]. The compiled .exe file will be called the same way as the one of RHSEG.

LA It is interesting to see how the aforementioned clustering methods perform. As a benchmark, it will also be considered how simple local averaging (LA) within rectangular windows performs. It is intuitively clear that this averaging will have a positive effect on the unmixing results in the case of homogeneous regions as eventually present noise will average out. However, the more important question is how it will influence the unmixing results in the case of merging pixels with spatially varying abundances and if more sophisticated clustering methods are needed in such a case, or if this simple method already provides acceptable results.

See Fig. 4.5 on how LA deals with window sizes that do not exactly fit in the image dimensions.

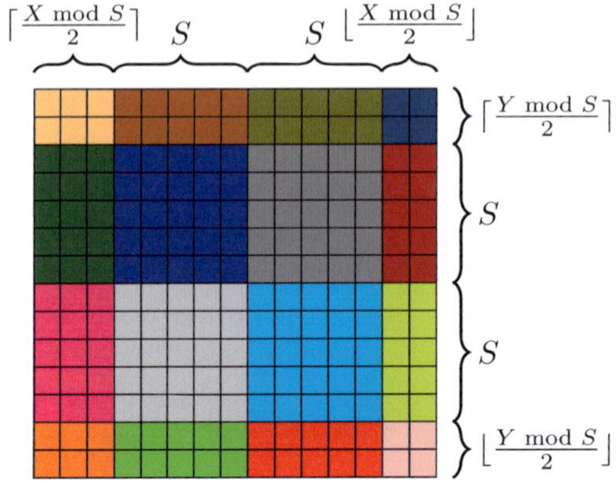

Figure 4.5 Local averaging method. In case the image dimensions X, Y are not divisible by the window size S without remainder, the image edges are treated as shown.

4.3 Pyramid Unmixing (PYRU)

In the previous Section 4.2, the UNCLI (UNmixing of CLustered Image) method describing the unmixing of clustered images has been presented. In anticipation of the UNCLI results that will be presented in Sec. 5.4, it can be stated that the local averaging (LA) method provides results that are not significantly worse than the ones obtained with sophisticated clustering methods. This motivates the application of LA to spectral unmixing in a slightly different way: remember the fact that the abundance update of the NMF objective function (3.1) is computationally much more demanding than the endmember update (this fact was already used in the UNCLI motivation). The new idea is to calculate endmembers and abundances using abundance approximations of lower spatial resolution in the first step, and use this coarse abundance approximation as initialization of the subsequent full resolution unmixing process. The low-resolution unmixing step can be split into several stages of increasing spatial resolution. Starting from a very low spatial resolution, this process can be described by lowpass image pyramids, see Fig. 4.6 [12, 71]. For this reason, the method is termed pyramid unmixing (PYRU). Although it is used here in the context of NMF, in general it is not limited to unmixing methods making use of NMF.

The main goal of PYRU is to speed up the calculation of the unmixing results, i.e., the same objective function value as obtained by regular unmixing is obtained in shorter time with PYRU. One factor that is beneficial for achieving this goal is the following one: in many cases, the abundance maps are initialized randomly, or by the constant $1/R$, or by the fully-constrained least-squares solution using the endmembers previously calculated by an endmember extraction algorithm. All these possibilities share the disadvantage that no spectral regularization has been taken into account for obtaining the abundances. By using PYRU, the spectral regularization can already be applied to the low-resolution levels, meaning that the resulting low-resolution abundance initialization of the full resolution unmixing step already has the regularization effect included. This means that most of the computational effort has been shifted to the low-resolution levels, where it can be accomplished faster, and optimally, only few optimization steps on the full resolution level are necessary.

4.3.1 Method Description

The PYRU procedure starts at the lowest spatial resolution image, i.e. the top one in Fig. 4.6, so let's assume here that it is $\mathbf{Y}^2$. After performing some unmixing steps on this level, the resulting abundance map is upsampled to the next higher resolution and unmixing is conducted on the next level image (one pyramid level below). Upsampling in this context means that all the pixels of the higher resolution image $\mathbf{Y}^1$ that have been used for creating one pixel of $\mathbf{Y}^2$ are assigned the abundance calculated for this pixel of $\mathbf{Y}^2$. After the unmixing calculations on level $\mathbf{Y}^1$ have been finished, the results are upsampled and the procedure continues on the next higher resolution image, in the considered example the full resolution image. Although image pyramids in general are based on a reduction factor of $N = 2$ which means $N \times N = 4$ pixels on the full resolution are averaged to obtain the next higher pyramid level, N in the context of PYRU can be chosen from powers of 2. Setting $N = 4$ means that every second pyramid level is skipped. The averaging can be described by the convolution of each band's 2D image with a square filter kernel of size $N \times N$ multiplied by N^2. Subsequently, downsampling by N is applied. Note that the spectral regularization parameter μ_{spec} is not adapted to each pyramid level; it is the same for all levels.

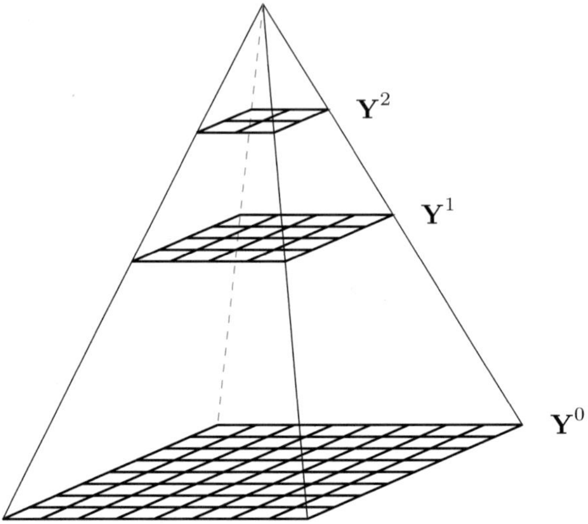

Figure 4.6 PYRU image pyramid; only one image band of each level is shown. The full resolution image is denoted by $\mathbf{Y}^0$, the lower resolution images are $\mathbf{Y}^1$ and $\mathbf{Y}^2$, respectively. These are obtained by averaging $N \times N$ pixels (here: $N = 2$) from the previous higher resolution image.

4.3.2 Implementation

When implementing the PYRU method, some practical aspects must be considered. The main point is how to efficiently perform the calculation of the low resolution images from the original one and the upsampling of the calculated abundance maps. The image pyramid is based on 2D images, while one 2D band of the hyperspectral image forms one line of the image matrix $\mathbf{Y}$. The information about neighboring pixels is lost in the matrix data representation.

In the first step, the dimension reduction of the image is performed. For each pyramid level, the downsampled image is stored in a 3D tensor

$\mathcal{Y}^K$. Also for each level, a two-dimensional so-called segmentation map $\mathbf{C}^k$ ($k = 1, \ldots, K$, where $K + 1$ is the total number of levels) is created and stored. It has the size of the previous (higher resolution) image and contains the information which pixels of the higher resolution image form one pixel of the lower resolution image. Then, the image at the highest pyramid level is converted to the matrix $\mathbf{Y}^K$ and the unmixing is performed, which yields the abundance maps $\mathbf{A}^K$. These are converted to the abundance cube $\mathcal{A}^K$ using the segmentation map, upsampled and the process continues on the next lower level. Figure 4.7 shows this procedure.

In case the size of the image to be downsampled is not divisible by N, the last original pixels form one downsampled pixel (this is by contrast to UNCLI, cf. Fig. 4.5). One example: when the size is 199 and $N = 4$, only the last 3 original pixels are considered for the last downsampled pixel. There is no cyclic or periodic usage of other image pixels.

For a beneficial application of PYRU, less iterations on the original resolution level should be performed than when only applying plain unmixing to the full image for saving calculation time. If this is not the case, the additional iterations on the lower-resolution levels that naturally have an additional time demand would make PYRU perform slower than plain unmixing. For selecting the number of iterations on each pyramid level, the formula

$$Z = \lceil V \cdot B^k \rceil \tag{4.25}$$

proved to be adequate. Z can be either P, M or T, i.e., the number of abundance iterations, the number of endmember iterations or the number of abundance/endmember sequences. $k = 0, \ldots, K$ denotes the pyramid level, where $k = 0$ denotes the original, full image resolution. Adapting the prefactor V and the base B, (4.25) enables choosing a large range of iteration numbers: For $B < 0$, less iterations are performed on the lower resolution levels than on the original image level. $B = 1$ leads to the same number of iterations on each level, while $B > 0$ performs more iterations on the lower resolution level than on full resolution. Note that when choosing $B < 0$, V should be chosen smaller than the number of iterations used for plain unmixing of the full resolution image for exploiting the previously mentioned PYRU time saving. It is reasonable to distinguish between the supervised and the unsupervised case for determining the iteration numbers:

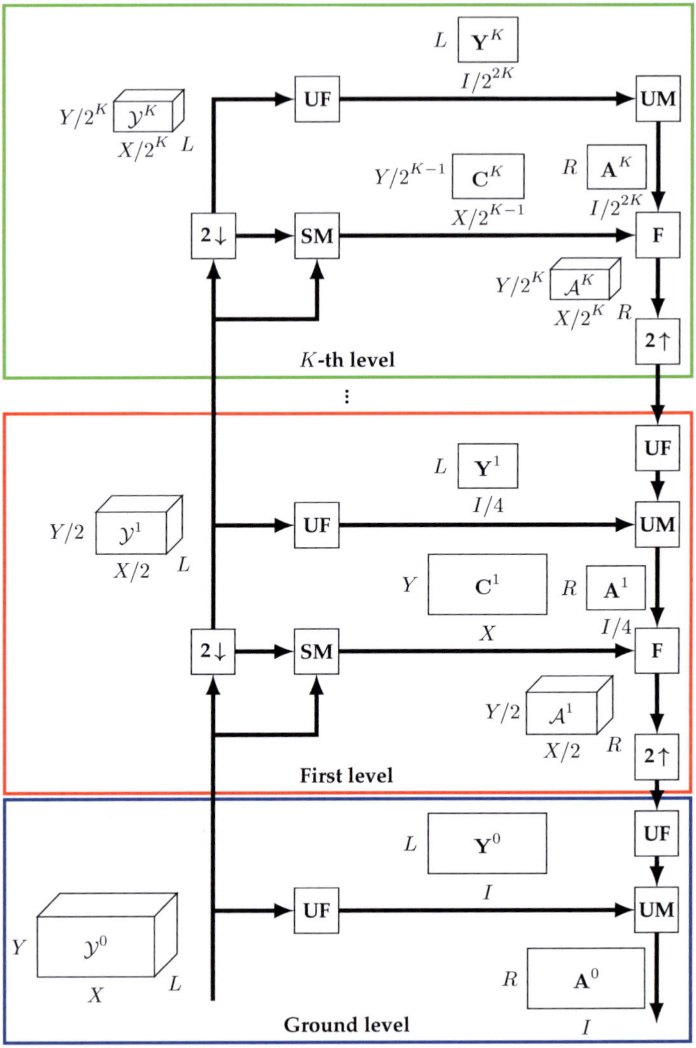

Figure 4.7 Pyramid unmixing procedure for $N = 2$. Upsampling and downsampling by a factor of 2 are performed by the blocks $2\uparrow$ and $2\downarrow$, respectively. The block SM computes the segmentation map, UM the unmixing, i.e., the abundance updates. UF (unfold) unfolds a tensor into a matrix, F (fold) vice versa.

- When the initial endmembers are chosen randomly from the image spectra (unsupervised case), P is constant and M is calculated using (4.25). T is also constant and chosen larger than one to allow for an alternating procedure between abundance and endmember updates.
- For the unsupervised case, it is also possible to adjust both P and M according to (4.25); on the low resolution levels, many endmember iterations can be performed, but the abundance calculation is of less importance, as the spatial resolution is inaccurate anyway. With increasing spatial resolution, the spatial accuracy becomes more important (increasing P), but the endmembers have been determined relatively precisely on the low resolution levels (decreasing M).
- A third alternative for the unsupervised case is the one used in the evaluation (Sec. 5.4). It consists of constant P and M on all levels, and adapting T according to (4.25).
- When the endmembers are determined by an endmember extraction algorithm (EEA) before unmixing (supervised case), M is set to one and only abundance iterations are performed while the endmembers remain unchanged. In this case, P is calculated according to (4.25). Another strategy which is also possible allows for subsequent endmember refinement during unmixing by setting $T > 1$, either fixed on each level or using (4.25).

Figure 4.8 shows the iteration numbers on each pyramid level in the unsupervised case.

Figure 4.9 depicts scatter plots (i.e., two bands plotted against each other) of two hyperspectral images. It illustrates how PYRU preserves the simplex structure (see Sec. 3.2) of the image pixels, i.e., except for the abundance accuracy requiring the full resolution, the spectral characteristics can already be fully exploited on the lower resolution levels.

4.4 3D Discrete Cosine Transform (DCT) Unmixing

The unmixing methods described so far have the major drawback that they are based on the data fidelity objective function $Q_{\text{data}}(\mathbf{M}, \mathbf{A}) = \|\mathbf{MA} - \mathbf{Y}\|_{\text{F}}^2$

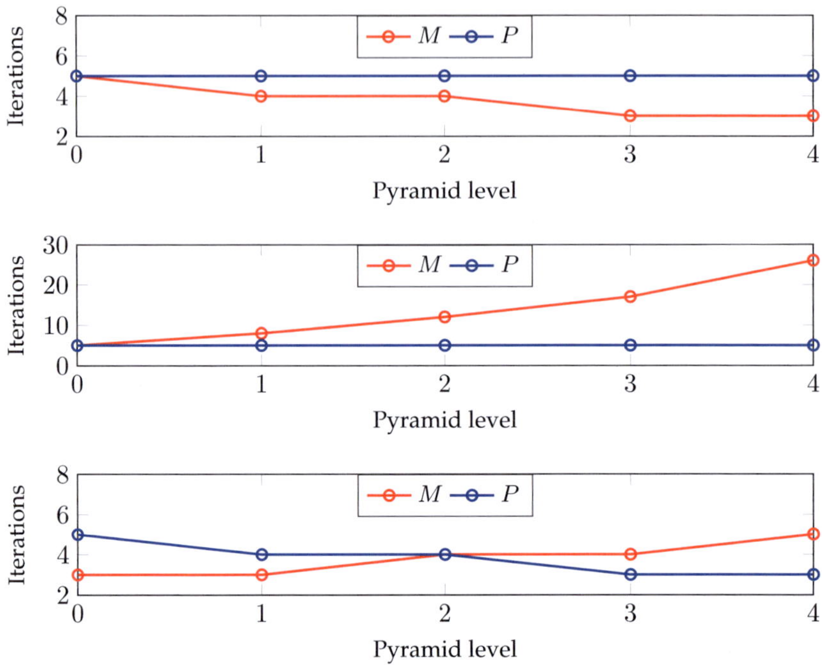

Figure 4.8 Exemplary calculation of the iteration numbers P and M according to (4.25) in the unsupervised case. Top graph: lower endmember iteration number on low resolution levels; middle graph: higher endmember iteration number on low resolution levels; bottom graph: high endmember iteration number on low resolution levels, high abundance iteration number on high resolution levels.

(3.1) which is biconvex, i.e., not convex for both $\mathbf{M}$ and $\mathbf{A}$ simultaneously, but convex in $\mathbf{M}$ for constant $\mathbf{A}$ and vice versa. This means that optimization based on gradient methods only can find local optima, but not the global minimum. In order for a function which is the sum of two functions to be convex, the two summands themselves have to be convex. This fact has the consequence that the objective function (3.2) with spectral and/or spatial regularization terms is non-convex, even if the regularizers are convex.

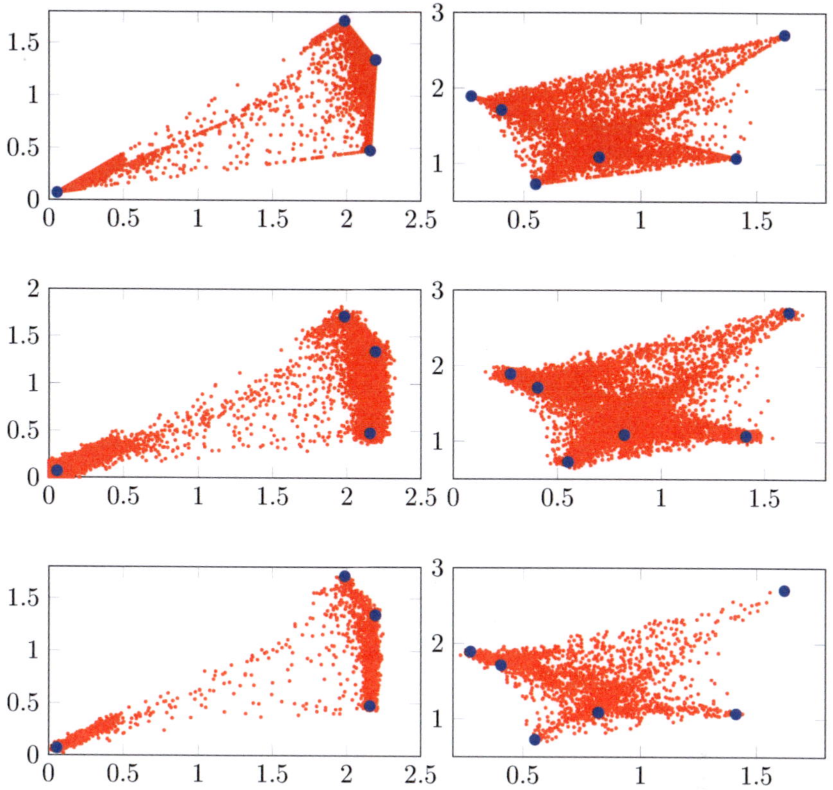

Figure 4.9 Left column: scatter plots of bands 80 (abscissa) and 180 (ordinate) of the Jasper image. Right column: scatter plots of bands 30 and 90 of the Urban Detail6 image. See Sec. 5.1.1 for explanations on the images. Each red dot represents an image pixel; the true endmembers are shown in blue. The top images are the original images, the second row shows the noisy images (SNR 30 dB) and the last row are the images on the first level of the pyramid with $N = 4$ (Jasper) and $N = 2$ (Urban Detail6). It can be seen that the averaging eliminates most of the noise, while the shape of the original simplex is well preserved, i.e., there are averaged image pixels very close to all true endmembers.

Another fact not considered in the discussed unmixing methods is that they do not consider global neighborhood information contained in the image values of the 3D hyperspectral tensor $\mathcal{Y}$. While TV and HS are based on local spatial pixel neighborhood information (remember Fig. 4.3), UNCLI can construct large non-local clusters incorporating many pixels, but mainly uses the spatial information. Even when spectral information is included, e.g., when k-means or RHSEG with its parameter $sw > 0$ are used, the full 3D correlations between image values of different pixels and different wavelengths are not used in a holistic way.

In this section, a new unmixing method providing a solution for both the non-convexity and the non-globality issue will be presented. Additionally, the method takes into account the fact that real-world signals can be represented in another transform domain with far less coefficients than in the original domain, while approximating the original signal with high fidelity or even losslessly [35]. This means for hyperspectral images that they can be represented approximately by the largest coefficients in some other transform domain. Utilizing the terminology from [35], this process is called *sparse approximation*. The new method is termed 3D discrete cosine transform nonnegative inverse matrix factorization (3DDCT-NIMF). The term "inverse" indicates that inverse matrices are used.

Figure 4.10 shows the transformation of a hyperspectral image tensor using the 3D discrete cosine transform (3D DCT) which will be used for the explained method. It is obvious that the coefficients corresponding to low frequencies f_x, f_y, f_l have a large magnitude in comparison with the other coefficients. By preserving only the high magnitude-coefficients and discarding the other ones, i.e., setting them to 0, the image is reconstructed with high quality after inverse 3D DCT transform. This fact is illustrated in Figs. 4.11 and 4.12 which demonstrate the approximation quality in case the lowest-magnitude coefficients are set to 0.

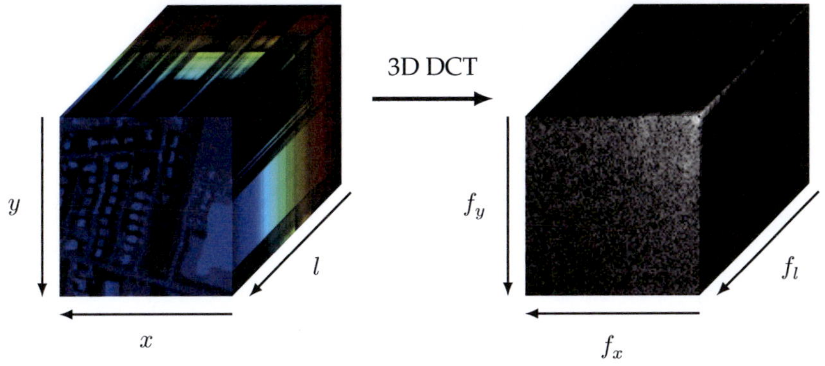

Figure 4.10 Data cube of the Urban Detail4 image (left, see Sec. 5.1.1) and data cube of its 3D DCT (right, logarithmic scale; mean of original image has been set to 0). Note that the main energy is concentrated around $f_x = f_y = f_l = 0$. Every element of the DCT cube depends on every element of the original cube. Interchanging two arbitrary, e.g., adjacent, elements of the image would affect the whole DCT cube.

4.4.1 Discrete Cosine Transform

The 1D DCT of a signal vector $\mathbf{u} \in \mathbb{R}^X$ is defined in the well-known publication [1] by

$$u_{\mathrm{DCT},f_x} = \begin{cases} \frac{\sqrt{2}}{X} \sum_{x=0}^{X-1} u_x & \text{for } f_x = 0 \\ \frac{2}{X} \sum_{x=0}^{X-1} u_x \cos\left(\frac{\pi}{X}\left(x + \frac{1}{2}\right) f_x\right) & \text{for } f_x = 1, 2, \ldots, X-1\,. \end{cases} \tag{4.26}$$

In this equation, X denotes the total number of elements in the signal, x the signal index, i.e., the index of the elements of $\mathbf{u}$, and f_x the frequency domain index of the DCT transformed vector $\mathbf{u}_{\mathrm{DCT}}$. This notation is chosen because it can easily be extended to additional signal dimensions. The transform can be thought of as the signal vector $\mathbf{u}$ multiplied by the

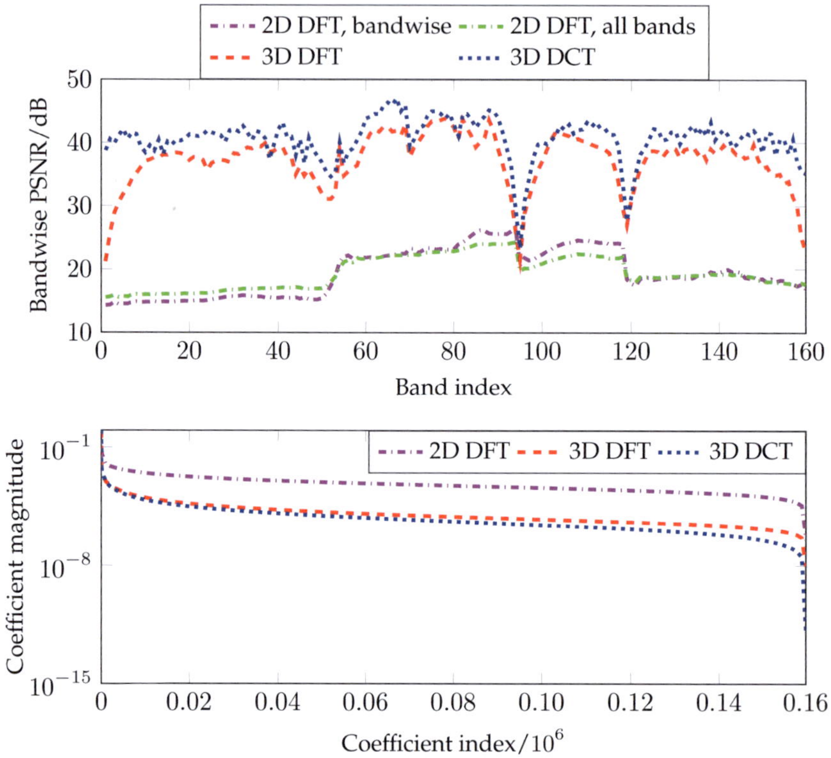

Figure 4.11 Urban Detail4 image analyzed with different transforms. Top plot: only 10 % of the largest transform coefficients are preserved, the rest is set to zero. Used transforms: bandwise 2D discrete Fourier transform (DFT) with the largest 10 % considered within each band (2D DFT, bandwise), and the coefficients of all bands jointly ordered and considering only the largest 10 % of this arrangement (2D DFT, all bands). Other transforms: 3D DFT and 3D DCT. Bottom plot: the coefficients are ordered by magnitude; all coefficients are normalized such that the largest one of each transform equals 1. The 3D DCT has the best compression properties: the top plot shows that it approximates the given image best; the bottom plot shows that all 3D DCT coefficients have smaller magnitude than the ones of the other transforms, meaning that the approximation error is smaller when they are disregarded, i.e., set to 0.

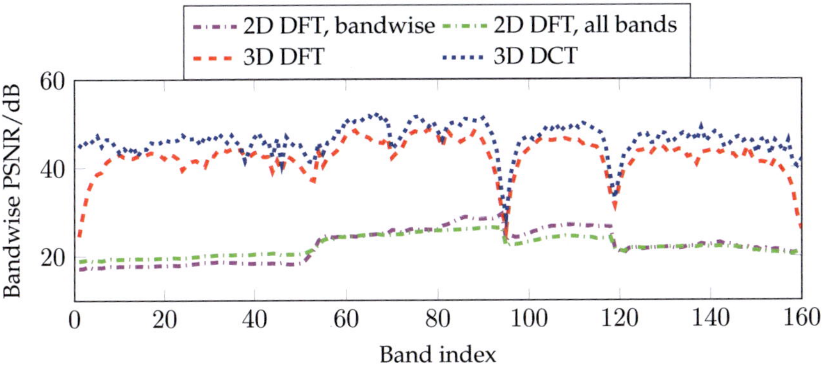

Figure 4.12 Same plot as the top one in Fig. 4.11, but this time, the largest 20 % of the coefficients are preserved. Note the higher PSNR values of each method compared to Fig. 4.11.

transform matrix $\mathbf{W}_X$,

$$\mathbf{u}_{\mathrm{DCT}} = \mathbf{W}_X \cdot \mathbf{u}\,, \tag{4.27}$$

where the transform matrix $\mathbf{W}_X \in \mathbb{R}^{X \times X}$ contains the cosine terms given in (4.26) in its respective elements. In order for the matrix to be orthogonal, these cosine terms are normalized a bit differently:

$$u_{\mathrm{DCT},f_x} = \begin{cases} \sqrt{\frac{1}{X}} \sum_{x=0}^{X-1} u_x & \text{for } f_x = 0 \\ \sqrt{\frac{2}{X}} \sum_{x=0}^{X-1} u_x \cos\left(\frac{\pi}{X}\left(x + \frac{1}{2}\right) f_x\right) & \text{for } f_x = 1, 2, \ldots, X-1\,. \end{cases} \tag{4.28}$$

Doing so has the consequence that the inverse transform matrix equals its transpose, $\mathbf{W}_X^{-1} = \mathbf{W}_X^{\mathrm{T}}$, as this is the definition of matrix orthogonality. As transform matrix orthogonality is a crucial point of the unmixing method that will be introduced shortly, the transform given in (4.28) will be used in the following. For brevity, the distinction between the vector element $u_{\mathrm{DCT},0}$ and all other elements arising from the different constants $\sqrt{\frac{1}{X}}$ and

$\sqrt{\frac{2}{X}}$ will be omitted for the remainder of this thesis. This means that the 1D DCT will be written

$$u_{\mathrm{DCT},f_x} = \frac{1}{\sqrt{X}} \sum_{x=0}^{X-1} u_x \cos\left(\frac{\pi}{X}\left(x+\frac{1}{2}\right) f_x\right) \tag{4.29}$$

for short, but having in mind that the prefactor value has to be considered in practice.

Using the matrix notation (4.27) and orthogonality, the inverse transform is given by

$$\mathbf{u}_{\mathrm{DCT}} = \mathbf{W}_X^{\mathrm{T}} \cdot \mathbf{u}\,. \tag{4.30}$$

For 2D images $\mathbf{U} \in \mathbb{R}^{X\times Y}$, it is possible to write the 2D DCT as matrix multiplications as well:

$$\mathbf{U}_{\mathrm{DCT}} = \mathbf{W}_X \cdot \mathbf{U} \cdot \mathbf{W}_Y\,, \tag{4.31}$$

where the first matrix $\mathbf{W}_X$ performs the 1D DCT along the columns of the image and the second matrix $\mathbf{W}_Y$ the 1D DCT along its rows.

The 3D DCT can be applied to the 3D hyperspectral image tensor $\mathcal{Y}$, resulting in the transformed tensor $\mathcal{Y}_{\mathrm{3DDCT}} \in \mathbb{R}^{X\times Y\times L}$, i.e., it has the same dimensions as the original one, see also Fig. 4.10. The elements of $\mathcal{Y}$ will be denoted by $y_{\mathbf{x}}$, where the image index vector is defined to $\mathbf{x} = [x, y, l]$. Analogously, the 3D DCT tensor $\mathcal{Y}_{\mathrm{3DDCT}}$ is indexed by the 3D DCT index vector $\mathbf{f} = [f_x, f_y, f_l]$, resulting in the tensor elements $y_{\mathrm{3DDCT},\mathbf{f}}$. Given this notation, the 3D DCT transform formula is defined by

$$y_{\mathrm{3DDCT},\mathbf{f}} = \frac{1}{\sqrt{XYL}} \sum_{x=0}^{X-1} \sum_{y=0}^{Y-1} \sum_{l=0}^{L-1} y_{\mathbf{x}} \cos\left(\frac{\pi}{X}\left(x+\frac{1}{2}\right) f_x\right) \cdot \cos\left(\frac{\pi}{Y}\left(y+\frac{1}{2}\right) f_y\right) \cos\left(\frac{\pi}{L}\left(l+\frac{1}{2}\right) f_l\right). \tag{4.32}$$

From this equation, it can be seen that each element $y_{\mathrm{3DDCT},\mathbf{f}}$ is calculated by summation over all image elements $y_{\mathbf{x}}$. This makes clear that using the 3D DCT, the full neighborhood information of all 3D image elements is exploited. In particular, it proves the fact stated in the caption of Fig. 4.10 that even simple interchanging of two adjacent image elements changes the whole 3D DCT cube.

4.4.2 3D DCT Applied to Spectral Unmixing

Remembering the fact that the hyperspectral image matrix $\mathbf{Y}$ is an unfolded version of the 3D tensor $\mathcal{Y}$, the 2D image $\mathbf{U}$ can be reshaped analogously such that the first line of pixels is positioned into the first Y elements of the horizontal vector $\mathbf{u}_{\text{image}} \in \mathbb{R}^{1 \times X \cdot Y}$, the second line into the second Y elements of this vector etc. In this case, it is necessary to have a single transform matrix $\mathbf{W}_{XY}$ performing the simultaneous 1D DCTs along X and Y, resulting in the matrix product

$$\mathbf{u}_{\text{image,DCT}} = \mathbf{u}_{\text{image}} \cdot \mathbf{W}_{XY} \,. \tag{4.33}$$

The matrix $\mathbf{W}_{XY}$ is designed such that the first Y elements of the transformed image vector $\mathbf{u}_{\text{image,DCT}}$ contain the frequencies f_y for constant $f_x = 0$, the next Y elements the frequencies f_y for constant $f_x = 1$ etc. This leads to the fact that $\mathbf{W}_{XY}$ is calculated by the Kronecker product

$$\mathbf{W}_{XY} = \mathbf{W}_X^{\text{T}} \otimes \mathbf{W}_Y \,. \tag{4.34}$$

For illustration, Fig. 4.13 gives a graphical overview of the steps described in the following. Starting point is the linear mixing model $\mathbf{Y} = \mathbf{MA}$ (2.2), step 1 in Fig. 4.13, where the noise and model error matrix $\mathbf{\Psi}$ is ignored for the moment. Transferring the principle of (4.33) to this mixing model and adding the third transform along the spectral dimension results in

$$\mathbf{W}_L \mathbf{Y} \mathbf{W}_{XY} = \mathbf{W}_L \mathbf{MAW}_{XY} \,, \tag{4.35}$$

see step 2 in Fig. 4.13. The columns of $\mathbf{Y}$ represent the image spectra, where the spectra of the first horizontal line of pixels of $\mathcal{Y}$ form the first Y columns, the second horizontal line the next Y columns etc. While the multiplication by $\mathbf{W}_{XY}$ performs the DCT in both spatial directions, the spectral DCT is realized by multiplication by the transform matrix $\mathbf{W}_L$.

The matrix orthogonality also holds for the transform matrix $\mathbf{W}_{XY}$,

$$\mathbf{W}_{XY} \cdot \mathbf{W}_{XY}^{\text{T}} = \mathbf{W}_{XY} \cdot \mathbf{W}_{XY}^{-1} = \mathbf{I} \,, \tag{4.36}$$

where the matrix $\mathbf{I}$ represents an identity matrix of adequate size.

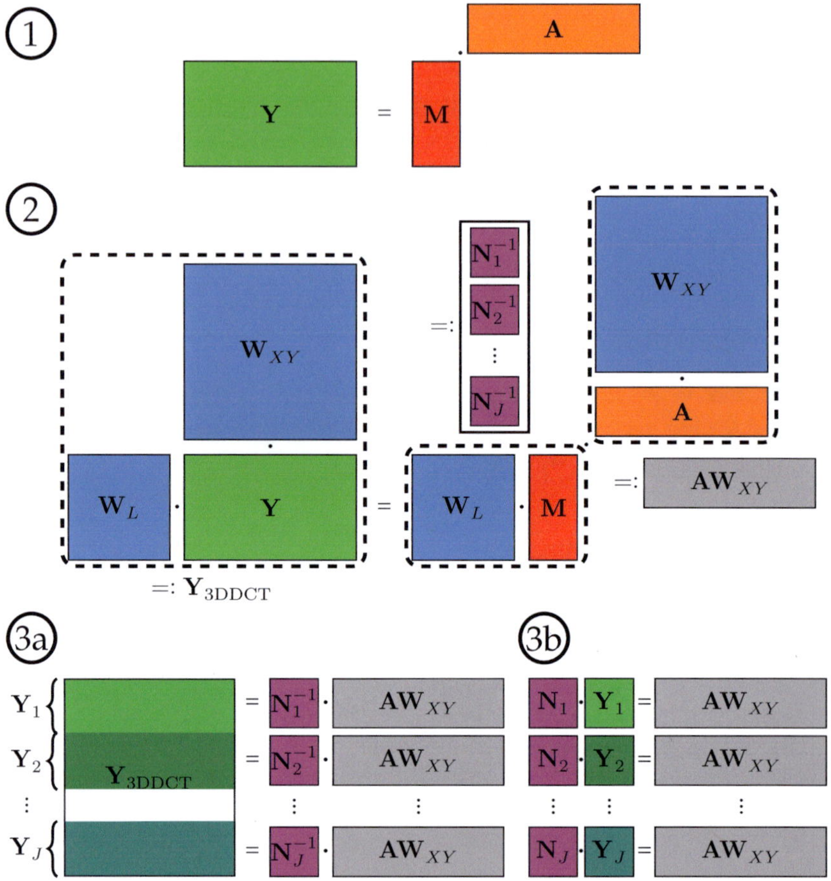

Figure 4.13 3DDCT-NIMF procedure. Step 1: linear mixing model used in conventional unmixing, $\mathbf{Y} = \mathbf{M} \cdot \mathbf{A}$. Applying the 3D DCT requires multiplying the 1D DCT matrix $\mathbf{W}_L$ from the left for performing the transform along the wavelength dimension. Multiplication from the right with $\mathbf{W}_{XY}$ realizes the two spatial DCTs (step 2). Step 3a shows the partitioning of the resulting matrices into blocks: the 3D DCT transformed image is split into submatrices of height R yiedling the submatrices $\mathbf{Y}_1, \ldots, \mathbf{Y}_J$ with $J = L/R$. In step 3b, the square matrices $\mathbf{N}_j^{-1}$, $j = 1, \ldots, J$, are inverted. Optimization yields the $\mathbf{N}_j$ and $\mathbf{A}$, see the objective function (4.41). After optimization, the endmember estimate $\mathbf{M}$ is determined from the calculated $\mathbf{N}_j$ by inverse 1D DCT as depicted in Fig. 4.14.

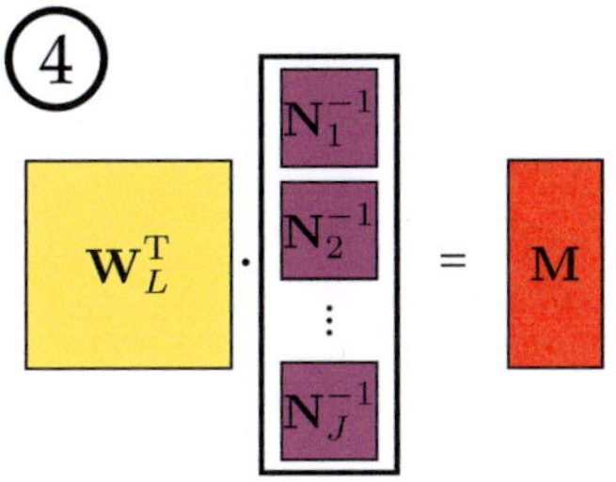

Figure 4.14 After optimization, the matrices $\mathbf{N}_j$ are inverted again and stacked on top of each other. Eventually, column by column IDCT (i.e., multiplication by $\mathbf{W}_L^{\mathrm{T}}$) is applied to this matrix. This multiplication yields the estimate of the endmember matrix $\mathbf{M}$ (final step 4 of 3DDCT-NIMF).

Due to the fact that both $\mathbf{W}_{XY}$ and $\mathbf{W}_L$ are orthogonal matrices, the objective function value of (2.2) does not change:

$$
\begin{aligned}
Q &= \|\mathbf{W}_L\mathbf{Y}\mathbf{W}_{XY} - \mathbf{W}_L\mathbf{M}\mathbf{A}\mathbf{W}_{XY}\|_{\mathrm{F}}^2 \qquad (4.37)\\
&= \mathrm{tr}\Big((\mathbf{W}_L\mathbf{Y}\mathbf{W}_{XY} - \mathbf{W}_L\mathbf{M}\mathbf{A}\mathbf{W}_{XY})(\mathbf{W}_L\mathbf{Y}\mathbf{W}_{XY} - \mathbf{W}_L\mathbf{M}\mathbf{A}\mathbf{W}_{XY})^{\mathrm{T}}\Big)\\
&= \mathrm{tr}\Big(\mathbf{W}_L\mathbf{Y}\mathbf{W}_{XY}\mathbf{W}_{XY}^{\mathrm{T}}\mathbf{Y}^{\mathrm{T}}\mathbf{W}_L^{\mathrm{T}} - 2\mathbf{W}_L\mathbf{Y}\mathbf{W}_{XY}\,\mathbf{W}_{XY}^{\mathrm{T}}\mathbf{A}^{\mathrm{T}}\mathbf{M}^{\mathrm{T}}\mathbf{W}_L^{\mathrm{T}}\\
&\quad + \mathbf{W}_L\mathbf{M}\mathbf{A}\mathbf{W}_{XY}\mathbf{W}_{XY}^{\mathrm{T}}\mathbf{A}^{\mathrm{T}}\mathbf{M}^{\mathrm{T}}\mathbf{W}_L^{\mathrm{T}}\Big)\\
&= \mathrm{tr}\Big(\mathbf{W}_L\mathbf{Y}\mathbf{Y}^{\mathrm{T}}\mathbf{W}_L^{\mathrm{T}} - 2\mathbf{W}_L\mathbf{Y}\mathbf{A}^{\mathrm{T}}\mathbf{M}^{\mathrm{T}}\mathbf{W}_L^{\mathrm{T}} + \mathbf{W}_L\mathbf{M}\mathbf{A}\mathbf{A}^{\mathrm{T}}\mathbf{M}^{\mathrm{T}}\mathbf{W}_L^{\mathrm{T}}\Big)\\
&= \mathrm{tr}\Big(\mathbf{W}_L^{\mathrm{T}}\mathbf{W}_L\mathbf{Y}\mathbf{Y}^{\mathrm{T}} - 2\mathbf{W}_L^{\mathrm{T}}\mathbf{W}_L\mathbf{Y}\mathbf{A}^{\mathrm{T}}\mathbf{M}^{\mathrm{T}} + \mathbf{W}_L^{\mathrm{T}}\mathbf{W}_L\mathbf{M}\mathbf{A}\mathbf{A}^{\mathrm{T}}\mathbf{M}^{\mathrm{T}}\Big)\\
&= \mathrm{tr}\Big(\mathbf{Y}\mathbf{Y}^{\mathrm{T}} - 2\mathbf{Y}\mathbf{A}^{\mathrm{T}}\mathbf{M}^{\mathrm{T}} + \mathbf{M}\mathbf{A}\mathbf{A}^{\mathrm{T}}\mathbf{M}^{\mathrm{T}}\Big)\\
&= \|\mathbf{Y} - \mathbf{M}\mathbf{A}\|_{\mathrm{F}}^2 .
\end{aligned}
$$

This means that after applying the 3D DCT, some additional operations should be performed in the DCT domain for having a difference to the common model. This can be done by incorporating sparse approximation: Only the ratio ρ of the largest absolute DCT coefficients will be preserved, while the other ones are set to 0. Mathematically, this can be expressed by

$$Q = \|\,(\mathbf{W}_L\mathbf{Y}\mathbf{W}_{XY}) \odot \mathbf{V} - (\mathbf{W}_L\mathbf{M}\mathbf{A}\mathbf{W}_{XY}) \odot \mathbf{V}\|_{\mathrm{F}}^2 . \qquad (4.38)$$

The operation $\odot$ denotes the Hadamard (elementwise) product, while $\mathbf{V} \in \mathbb{R}^{L \times I}$ is a filter matrix of the same size as $\mathbf{Y}$ that consists of zeros and ones, therefore realizing the sparse approximation. Applying the same calculation as in (4.37) also to (4.38) makes it obvious that the filter matrix is located between the transform matrices and its respective transpose. This means that no simplification can be performed, and (4.38) does not convert to the original objective function $\mathbf{Y} = \mathbf{MA}$.

Instead of using the expression with a filter matrix, there is a more elegant way of mathematically representing the sparse approximation. The transform matrix $\mathbf{W}_{XY}$ performs the two spatial DCTs of $\mathbf{Y}$ and the abundance matrix $\mathbf{A}$, meaning that it contains in its columns the cosine terms required for calculating the frequencies f_x and f_y. The first column contains the cosine terms of the frequencies $f_x = 0$, $f_y = 0$ and so on. If a frequency pair (f_{x_0}, f_{y_0}) is not contained in all rows of the 3D DCT transformed image matrix, i.e., all elements of this respective column of the 3D DCT transformed image matrix have been set to zero for performing sparse approximation, the correspondent column of $\mathbf{W}_{XY}$ can be deleted, so that this frequency pair of $\mathbf{A}$ is not calculated at all. As a consequence, $\mathbf{W}_{XY}$ is not an orthogonal matrix anymore, and the simplification (4.37) is no longer valid. Interestingly, when taking the inverse transform also into account, the product $\mathbf{W}_{XY}\mathbf{W}_{XY}^{\mathrm{T}}$ is in fact equivalent to calculating the 2D DCT, zeroing the respective frequency coefficients and calculating the inverse DCT. This is based on the fact that it does not matter for the inverse transform if all frequencies are calculated during the 2D DCT, and the respective ones zeroed afterwards, or the respective frequencies have been left out from the beginning.

For calculating the abundance maps, this is a reasonable way: if the acquired hyperspectral image (after zeroing the smaller coefficients) does not contain a spatial frequency pair, the calculated abundance matrices also should not contain this frequency pair. By deleting the correspondent column of $\mathbf{W}_{XY}$, the implicit constraint that this frequency pair is set to zero in the abundances is incorporated. Note that the spectral transform matrix $\mathbf{W}_L$ remains unchanged; all frequencies f_l will be considered.

4.4.3 Objective Function Convexification

So far, the 3D DCT objective function (4.38), or to be more precise, its realization with adapted transform matrices $\mathbf{W}_{XY}$, fulfills the requirement that all spectral and spatial correlations contained in the full data cube are incorporated. Additionally, sparse approximation is integrated. Unfortunately, the resulting objective function (4.38) is still not convex. For convexifying it, an approach already used in, e.g., [6, 161] is used. In contrast to these references, where the data matrix is reduced to one single matrix by PCA, the 3D DCT transformed matrix $\mathbf{Y}$ is split into horizontal blocks of size $R \times I$ (disregarding superfluous bands if there is a remainder after dividing L by R), see Fig. 4.13. The main point, however, is that the references assume the noise-free case, while the presented 3DDCT-NIMF method is designed explicitly for noisy images by exploiting the resulting SNR properties that will be explained in Sec. 4.4.4.

The 3DDCT-NIMF method first calculates the 3D DCT transformed image matrix $\mathbf{Y}_{3\text{DDCT}} := \mathbf{W}_L \mathbf{Y} \mathbf{W}_{XY}$ that is subsequently split into $J = L/R$ horizontal blocks $\mathbf{Y}_1, \ldots, \mathbf{Y}_J$ (step 3a in Fig. 4.13). The result of the matrix product $\mathbf{W}_L \cdot \mathbf{M}$ is also split into matrices $\mathbf{M}_j$ which are of dimension $R \times R$. As the $\mathbf{M}_j$ are assumed to be nonsingular square matrices, they can be inverted: $\mathbf{N}_j := \mathbf{M}_j^{-1}$. This is depicted in step 3b of Fig. 4.13. In total, (4.35) is split into blocks

$$\mathbf{N}_j \left(\mathbf{W}_L \mathbf{Y} \mathbf{W}_{XY} \right)_j = \mathbf{A} \mathbf{W}_{XY,j} \tag{4.39}$$

and the objective function converts to

$$Q_{3\text{DDCT}} = \sum_{j=1}^{J} \| \mathbf{N}_j \mathbf{Y}_j - \mathbf{A} \mathbf{W}_{XY,j} \|_{\text{F}}^2 \, . \tag{4.40}$$

In this equation, the matrices $\mathbf{W}_{XY}$ were given indices to point out that the transform matrix is adapted for each block j. The reason is that depending on the correspondent matrices $\mathbf{Y}_j$, $j = 1, \ldots, J$, i.e., on the frequency content in the respective image block, more or less frequencies are left out. The first block $\mathbf{Y}_1$ contains the lowest frequencies f_l, while the next ones contain higher frequencies. As illustrated in Fig. 4.10, higher frequencies have smaller coefficients, which means that with increasing block index j, more coefficients are likely to be set to zero. This is taken into account by adapting $\mathbf{W}_{XY,j}$ depending on the block number.

The resulting objective function Q_{3DDCT} (4.40) is convex. This is due to the fact that compositions of an affine mapping and a convex function are convex [19, Sec. 3.2.2]: for convex $Q : \mathbb{R}^n \to \mathbb{R}$, $\mathbf{C} \in \mathbb{R}^{n \times m}$, $\mathbf{d} \in \mathbb{R}^n$, $Q(\mathbf{Cx} + \mathbf{d})$ is convex as well. All matrix norms are convex, therefore also the Frobenius norm, and all elements of the matrices $\mathbf{N}_j$ and $\mathbf{A}$ can be rearranged to one large vector. This resulting vector is multiplied by a matrix containing constant elements resulting from, e.g., the transform matrices $\mathbf{W}_{XY,j}$ and the 3D DCT image matrices $\mathbf{Y}_j$. This is an affine mapping. Combining all these facts, (4.40) is convex.

Unfortunately, the optimal solution of (4.40) are the matrices $\mathbf{N}_j = \mathbf{0}$, $\mathbf{A} = \mathbf{0}$; the Frobenius norm has minimum value 0, which is obtained for exactly these matrices. For this reason, an additional constraint or a regularizer has to be added for preventing the matrices from becoming $\mathbf{0}$. Just like in for instance [6], the term $-\log(|\det(\mathbf{N}_j)|)$ is used for this purpose. It is derived from maximum likelihood considerations [65, Secs. 2.6, 9.1.1]. This regularizer is non-convex. If the $\mathbf{N}_j$ would be restricted to be positive definite, the term $-\log(\det(\mathbf{N}_j))$ without the absolute would be convex [19, Sec. 3.1.5]. This, however, would be too restrictive as the only requirement is that the $\mathbf{N}_j$ are regular, i.e., invertible, and not $\mathbf{0}$. Adding the regularizer yields the objective function

$$Q_{\text{3DDCT}} = \sum_{j=1}^{J} \|\mathbf{N}_j \mathbf{Y}_j - \mathbf{A}\mathbf{W}_{XY,j}\|_{\text{F}}^2 - K_j \log(|\det(\mathbf{N}_j)|) \,. \tag{4.41}$$

The K_j are block-dependent regularization parameters balancing the 3D DCT data fidelity term and the regularization term ensuring that $\mathbf{N}_j$ is regular. They are calculated by the heuristic formula

$$K_j = K \cdot \left(P \frac{\sum_{r=1}^{R} \sum_{i=1}^{I} |\mathbf{Y}_{j,ri}|}{\text{nnz}(\mathbf{Y}_j)} \right)^E \,. \tag{4.42}$$

This formula proved to be useful for all experiments; it reduces the number of parameters from J to just two, while providing good unmixing results. The remaining two parameters are $E > 0$ for including some non-linearity, while $K > 0$ is constant for all J blocks. In (4.42), $\mathbf{Y}_{j,ri}$ denotes the element of $\mathbf{Y}_j$ in the r-th row and the i-th column. The operation $\mathrm{nnz}(\cdot)$ provides the number of nonzero matrix elements. As the sparse approximation already had been applied at this point in time, many matrix elements are 0. The purpose of the constant $P \gg 1$ is to ensure that the term within the brackets of (4.42) is > 1. Altogether, the parameters K_j are obtained by calculating the average of the nonzero elements of the respective matrix $\mathbf{Y}_j$, multiplying by the large constant P, exponentiating by E and multiplying by the constant K. The value of P is constant for all considerations and evaluations in this thesis.

After minimizing (4.41), the inverses of all matrices $\mathbf{N}_j$ are stacked on top of each other. Column by column inverse 1D DCT of the resulting matrix provides the final matrix $\mathbf{M}$ (see also Fig. 4.14).

The gradients of (4.41) are given by

$$\frac{\partial Q_{\mathrm{3DDCT}}}{\partial \mathbf{A}} = \sum_{j=1}^{J} 2\mathbf{A}\mathbf{W}_{XY,j}\mathbf{W}_{XY,j}^{\mathrm{T}} - 2\mathbf{N}_j\mathbf{Y}_j\mathbf{W}_{XY,j}^{\mathrm{T}}\,, \tag{4.43}$$

$$\frac{\partial Q_{\mathrm{3DDCT}}}{\partial \mathbf{N}_j} = 2\mathbf{N}_j\mathbf{Y}_j\mathbf{Y}_j^{\mathrm{T}} - 2\mathbf{A}\mathbf{W}_{XY,j}\mathbf{Y}_j^{\mathrm{T}} - \left(\mathbf{N}_j^{\mathrm{T}}\right)^{-1}\,. \tag{4.44}$$

These are required for applying the projected quasi-Newton (PQN) method which will be explained in Sec. 4.6.1. Because PQN is implemented in MATLAB and uses function handles, it is possible to circumvent the need for performing explicit calculations with the transform matrices $\mathbf{W}_{XY,j}$. This is advantageous as they become very large due to the Kronecker product (4.34): for $X = Y = 100$, $\mathbf{W}_{XY,j}$ has the size 10000×10000 if all frequencies are incorporated. Instead, e.g., for implementing the product $\mathbf{A}\mathbf{W}_{xy,j}\mathbf{W}_{xy,j}^{\mathrm{T}}$ from (4.43), it is possible to reshape the rows of $\mathbf{A}$ (length I) back to images of size $X \times Y$ and straightforwardly apply the direct 2D transform version instead of the one derived for directly using the (unfolded) rows of $\mathbf{A}$ that is contained in $\mathbf{W}_{XY,j}$.

Although the gradients (4.43) and (4.44) will be used for the remainder of this thesis, it might be necessary for future adaptions of 3DDCT-NIMF to include weighting in the frequency domain, e.g., putting more weight on higher frequency coefficients. This means that the objective function

(4.40) contains weighting matrices $\mathbf{V}$ similarly to (4.38):

$$Q = \sum_{j=1}^{J} \|(\mathbf{N}_j \mathbf{Y}_j) \odot \mathbf{V}_j - (\mathbf{A}\mathbf{W}_{xy,j}) \odot \mathbf{V}_j\|_{\mathrm{F}}^2 . \tag{4.45}$$

Now, the matrix $\mathbf{V}$ does not only contain zeros and ones, but also other values. For this objective function, the calculation of the gradients is given in Appendix A.

Technically, it would be possible to directly split the LMM (2.2) into blocks and apply blockwise matrix inversion. There are, however, two main problems connected with this approach. First, the calculated endmembers are discontinuous, because each matrix $\mathbf{N}_j$ would be inverted separately, meaning that there is no connection between adjacent blocks. Second, if noise is present in the image matrix $\mathbf{Y}$, the optimization would yield matrices $\mathbf{N}_j$ that deviate from the true matrices. Matrix inversion is highly sensitive towards deviations in the original elements, which leads to the fact that even small misestimations can greatly falsify the inverted matrices, therefore leading to very inaccurate endmember estimation. Due to the fact that 3DDCT-NIMF includes the DCT along the spectral direction, the first problem is circumvented in an elegant way, because each block j estimates some of the DCT frequencies of the endmember matrix $\mathbf{M}$. The full matrix $\mathbf{M}$ is obtained after inverse DCT, which is equivalent to the weighted addition of cosine basis functions. Because each of these basis functions is continuous, the estimated spectra also will be continuous. How 3DDCT-NIMF is dealing with data corrupted by noise is subject of the next section.

4.4.4 Characteristics of 3DDCT Nonnegative Inverse Matrix Factorization (NIMF) in the Presence of Noise

In Sec. 4.4.2, a convexification step has been applied to the linear mixing model (2.2), leading to a convex objective function with non-convex regularization term. At that time, the noise and model error matrix $\mathbf{\Psi}$ was ignored. Considering this matrix leads to some interesting insights about the properties of 3DDCT-NIMF regarding the noise contained in each frequency coefficient.

Applying the 3D DCT to $\mathbf{\Psi}$ yields the matrix

$$\mathbf{\Psi}_{\text{3DDCT}} = \mathbf{W}_L \mathbf{\Psi} \mathbf{W}_{XY} \,. \tag{4.46}$$

Due to the fact that noise contribution and model error are unknown, $\mathbf{\Psi}$ represents a realization of a matrix of random variables. This means that in a general model description, applying the 3D DCT multiplies all random variables by cosine terms and sums them up. It is assumed that all random variables are independent, which leads to the conclusion that, according to the central limit theorem, the resulting random variables are normally distributed regardless of the distribution of the single original random variables. For this reason, one does not need to use distances other than the Frobenius norm that would be required for different noise distributions [31, Ch. 2]. The least-squares distance, i.e., the Frobenius norm, is optimal in the presence of normally distributed uncorrelated noise (Sec. 2.2.2). Having determined the distribution type of the noise corrupting the frequency coefficients, the next question is which variance the noise normal distributions of each coefficient have.

As explained in Sec. 2.2.5, adding random variables is equivalent to multiplying their characteristic functions. Considering arbitrary time-dependent functions $g_1(t)$, $g_2(t)$ and $g_3(t)$ and their respective frequency-domain correspondences $G_1(f)$, $G_2(f)$ and $G_3(f)$ leads to the Fourier correspondence

$$g_1(t) = g_2(t) * g_3(t) \;\circ\!\!-\!\!\bullet\; G_1(f) = G_2(f) \cdot G_3(f) \,. \tag{4.47}$$

In this context, note that the functions $G_2(f)$ and $G_3(f)$ have the same argument as the calculated function $G_1(f)$.

One notational convention used in the following is that outcomes of random variables (e.g., z) will be written lowercase. The corresponding variables in the Fourier domain will be written uppercase. Another convention is that the characteristic functions of random variables are denoted by capital $F(\cdot)$.

The following Fourier correspondence between density and characteristic function holds for the normally distributed random variable z (2.18):

$$\frac{1}{\sigma\sqrt{2\pi}} \mathrm{e}^{-\frac{z^2}{2\sigma^2}} \;\circ\!\!-\!\!\bullet\; \mathrm{e}^{-2\pi^2\sigma^2 Z^2} \,. \tag{4.48}$$

Using the aforementioned convention, Z denotes the frequency domain variable corresponding to z. Using the scaling of random variables (2.17) leads to the correspondence

$$\frac{1}{|k|}\frac{1}{\sigma\sqrt{2\pi}}\mathrm{e}^{-\frac{z^2}{2\sigma^2 k^2}} \circ\!\!-\!\!\bullet\ \mathrm{e}^{-2\pi^2\sigma^2 k^2 Z^2}\,. \tag{4.49}$$

As mentioned before, all image values of the image tensor $\mathcal{Y}$ have a noise random variable whose realizations are collected in the (unfolded) matrix $\mathbf{\Psi}$. The noise random variable at position $\mathbf{x} = [x, y, l]$ of $\mathcal{Y}$ is denoted $\mathsf{z}_{\mathbf{x}}$. The variables at all positions are assumed to be normally distributed with variance σ^2. Application of the 3D DCT leads to the random variables $\mathsf{Z}_{\mathbf{f}}$ at positions $\mathbf{f} = [f_x, f_y, f_l]$ of the noise tensor in the DCT domain:

$$\mathsf{Z}_{\mathbf{f}} = \sum_{x=0}^{X-1}\sum_{y=0}^{Y-1}\sum_{l=0}^{L-1} \mathsf{z}_{\mathbf{x}} c_{\mathbf{x},\mathbf{f}}\,. \tag{4.50}$$

These random variables $\mathsf{Z}_{\mathbf{f}}$ describe the noise in each frequency coefficient. Due to the fact that the DCT basis functions can be considered as constants for the following train of thoughts, they are given a shorter notation:

$$\begin{aligned} c_{\mathbf{x},\mathbf{f}} :=& \frac{1}{\sqrt{XYL}}\cos\left(\frac{\pi}{X}\left(x+\frac{1}{2}\right)f_x\right) \\ &\cdot\cos\left(\frac{\pi}{Y}\left(y+\frac{1}{2}\right)f_y\right)\cos\left(\frac{\pi}{L}\left(l+\frac{1}{2}\right)f_l\right)\,. \end{aligned} \tag{4.51}$$

The densities of the scaled random variables $\mathsf{z}_{\mathbf{x},c} := \mathsf{z}_{\mathbf{x}} c_{\mathbf{x},\mathbf{f}}$ in (4.50) are given by

$$f_{\mathsf{z}_{\mathbf{x},c}}(z_{\mathbf{x},c}) = \frac{1}{|c_{\mathbf{x},\mathbf{f}}|}\frac{1}{\sigma\sqrt{2\pi}}\mathrm{e}^{-\frac{z_{\mathbf{x},c}^2}{2\sigma^2 c_{\mathbf{x},\mathbf{f}}^2}}\,. \tag{4.52}$$

The characteristic functions are obtained by Fourier transform of (4.52):

$$F_{\mathsf{Z}_{\mathbf{x},c}}(Z_{\mathbf{x},c}) = \mathrm{e}^{-2\pi^2\sigma^2 c_{\mathbf{x},\mathbf{f}}^2 Z_{\mathbf{x},c}^2}\,. \tag{4.53}$$

Following (4.50), the random variables have to be added. This means that the respective densities $f_{\mathsf{Z}_{\mathbf{f}}}(z_{\mathbf{f}})$ are calculated by convolution of all the densities $f_{\mathbf{Z}_{\mathbf{x},c}}(z_{\mathbf{x},c})$ for $x = 0, \ldots, X-1, y = 0, \ldots, Y-1, l = 0, \ldots, L-1$. This is equivalent to multiplying the characteristic functions in the Fourier domain:

$$F_{\mathbf{Z}_\mathbf{f}}(Z_\mathbf{f}) = \prod_{x=0}^{X-1}\prod_{y=0}^{Y-1}\prod_{l=0}^{L-1} F_{\mathbf{Z}_{\mathbf{x},c}}(Z_\mathbf{f}) = \prod_{x=0}^{X-1}\prod_{y=0}^{Y-1}\prod_{l=0}^{L-1} \mathrm{e}^{-2\pi^2\sigma^2 c_{\mathbf{x},\mathbf{f}}^2 Z_\mathbf{f}^2}\,. \tag{4.54}$$

The crucial point is that using (4.47), the arguments of all $F_{\mathbf{Z}_{\mathbf{x},c}}(Z_{\mathbf{x},c})$ need to be changed from $Z_{\mathbf{x},c}$ to $Z_\mathbf{f}$ because the sought-after function $F_{\mathbf{Z}_\mathbf{f}}(Z_\mathbf{f})$ carries this argument. Further simplification of (4.54) yields

$$F_{\mathbf{Z}_\mathbf{f}}(Z_\mathbf{f}) = \exp\left(-2\pi^2\sigma^2 Z_\mathbf{f}^2 \sum_{x=0}^{X-1}\sum_{y=0}^{Y-1}\sum_{l=0}^{L-1} c_{\mathbf{x},\mathbf{f}}^2\right), \tag{4.55}$$

because $Z_\mathbf{f}^2$ is independent of x, y and l. Taking a closer look at the sum over the squared constants, one obtains

$$\sum_{x=0}^{X-1}\sum_{y=0}^{Y-1}\sum_{l=0}^{L-1} c_{\mathbf{x},\mathbf{f}}^2 = \frac{1}{X}\sum_{x=0}^{X-1}\cos^2\left(\frac{\pi}{X}\left(x+\frac{1}{2}\right)f_x\right) \tag{4.56}$$
$$\cdot \frac{1}{Y}\sum_{y=0}^{Y-1}\cos^2\left(\frac{\pi}{Y}\left(y+\frac{1}{2}\right)f_y\right)\underbrace{\frac{1}{L}\sum_{l=0}^{L-1}\cos^2\left(\frac{\pi}{L}\left(l+\frac{1}{2}\right)f_l\right)}_{=1}.$$

This is due to the fact that the DCT transform matrices (see Sec. 4.4.1) are orthogonal. The underbraced part of (4.55) can be thought of as the dot product of one row vector of the transform matrix $\mathbf{W}_L$ with itself which results to 1. Treating the two sums over x and y the same way leads to the simple equation

$$\sum_{x=0}^{X-1}\sum_{y=0}^{Y-1}\sum_{l=0}^{L-1} c_{\mathbf{x},\mathbf{f}}^2 = 1\,. \tag{4.57}$$

It is important to remark that this holds for all frequencies f_x, f_y and f_l which provides a simple version of (4.55):

$$F_{\mathbf{Z}_\mathbf{f}}(Z_\mathbf{f}) = \mathrm{e}^{-2\pi^2\sigma^2 Z_\mathbf{f}^2}\,. \tag{4.58}$$

Using again the correspondence (4.48) provides the noise densities of all frequency coefficients

$$f_{\mathbf{z_f}}(z_{\mathbf{f}}) = \frac{1}{\sigma\sqrt{2\pi}} \mathrm{e}^{-\frac{z_{\mathbf{f}}^2}{2\sigma^2}} . \tag{4.59}$$

The meaning of this equation is that the variance of the noise in all frequency coefficients is the same and equals the one of the variance in all elements of the original image $\mathcal{Y}$. As a consequence, the SNR in the image and the DCT domain is the same.

The most interesting fact about the derivation above becomes obvious when considering the signal energy scattered in the 3D DCT transformed data cube. As already discussed and illustrated in Fig. 4.10, the coefficients of the lowest frequencies have the highest magnitude, while high-frequency coefficients have a far lower magnitude. This is confirmed mathematically by the so-called Riemann-Lebesgue lemma [136]. As a consequence, the SNR is different in different regions of the 3D DCT cube: the lowest frequencies have the highest SNR, while higher frequencies have lower SNR. This is shown in Fig. 4.15.

The SNR properties of the DCT cube can be exploited beneficially for 3DDCT-NIMF. The calculation of the unmixing is performed in the first step using all blocks of the objective function (4.41). For the last minimization step of $\mathbf{A}$, only the first block ($j = 1$) is considered, because this block contains the highest SNR, and the full abundance matrix $\mathbf{A}$ is already contained in it. The other blocks have far lower SNR and merely contain redundant information. Having calculated the final matrix $\mathbf{A}$, the final matrices $\mathbf{N}_j$ ($j = 1, \ldots J$) are calculated using this final $\mathbf{A}$, but this time only incorporating the spatial frequencies $f_x, f_y = 0, \ldots, B - 1$, where $B \leq \min\{X, Y\}$. These are coefficients with high SNR, see the bright band in direction of f_l at low spatial frequencies in the top graphic of Fig. 4.15. Technically, the consideration of only the low spatial frequencies can be realized by further adapting the transform matrices $\mathbf{W}_{XY,j}$ and deleting the respective columns of these higher frequencies. From these considerations, graphically illustrated in Fig. 4.15, it can be seen that considering only the mentioned frequency coefficients increases the robustness of 3DDCT-NIMF to noise. The average SNR in the incorporated coefficients is much higher than the average SNR in the original image.

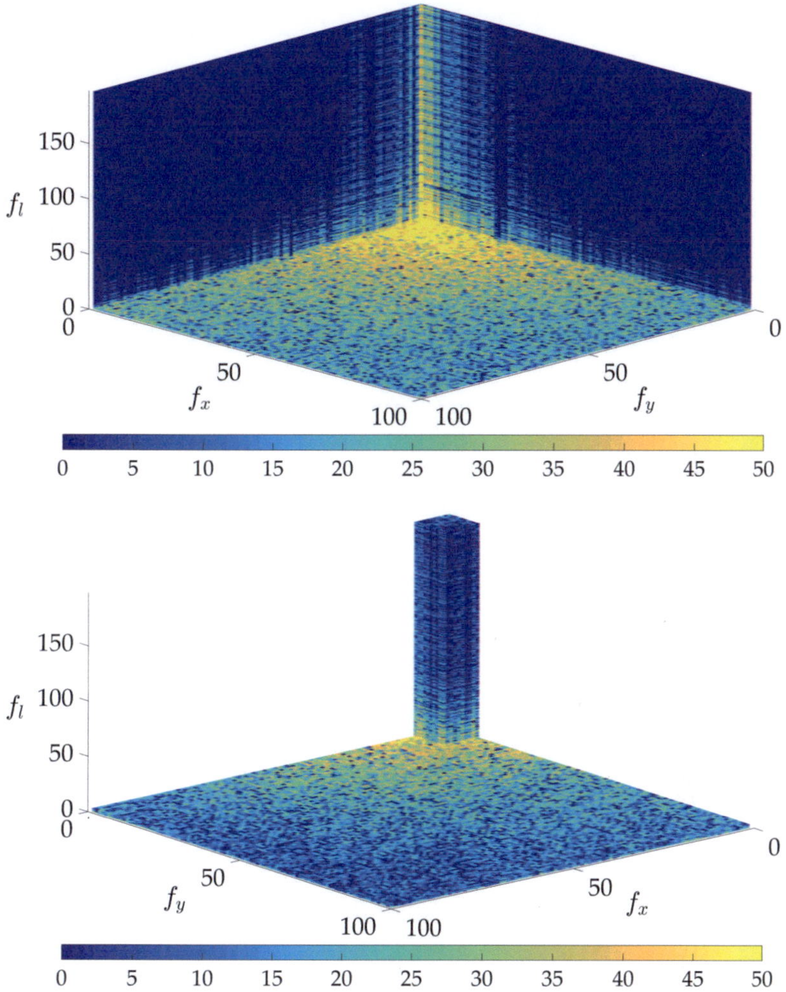

Figure 4.15 Jasper image (Sec. 5.1.1): DCT SNR analysis with 30 dB overall SNR in the original image. Top: DCT coefficient-wise SNR in dB, color encoded. As the cube is shown from a perspective below it, the bright corner comes out of the page. Bottom: Same cube from a different angle, this time considering only the first block ($j = 1$) and 10×10 coefficients in the other blocks. Average SNR of these coefficients: 45.22 dB, much higher than total average of 30 dB. The color scaling of both images has been cut off to allow for better visual perception.

4.4.5 Resolving the Scaling Ambiguity

Determining the elements of an inverse matrix is one major way to compute an independent component analysis (ICA) decomposition [65, Sec. 1.3.1], for instance by minimization of the mutual information [65, Ch. 10]. The ICA decomposition is unique except for permutation and scaling [65, Sec. 7.2.3]. This is also the case with 3DDCT-NIMF, however, the sum-to-one constraint (STO constraint) provides a possibility to solve the scaling ambiguity. Permutation is not of significance, as the order of the abundance maps and the endmembers is not important. It is only required for determination of the decomposition quality, and in this case can be resolved by comparing the ground truth and the calculated abundances and ordering the calculated ones accordingly.

3DDCT-NIMF uses the regularization parameters K_j, see the objective function (4.41). Their impact is that they change the scaling of the matrices $\mathbf{N}_j$ and $\mathbf{A}$: for larger K_j, the determinant is increased, leading to larger absolute values of the elements of $\mathbf{N}_j$. Subsequently, in order to keep the Frobenius norm (data fidelity) in the objective function small, the values of the elements of $\mathbf{A}$ also have to increase. The matrices $\mathbf{N}_j$ can have negative values due to the fact that they have to account for DCT and inversion, whereas the elements of $\mathbf{A}$ are restricted to be nonnegative. For this reason, the elements of $\mathbf{A}$ can be used together with the STO constraint for solving the scaling ambiguity: after $\mathbf{A}$ has been calculated from the highest SNR block of the DCT cube (see end of previous section), its columns (i.e., the abundances of each pixel) are normalized to sum to one. This is performed by dividing each column by the sum of its elements. In the next step, the matrices $\mathbf{N}_j$ are used to calculate the endmember matrix $\mathbf{M}$. Using the normalized matrix $\mathbf{A}$, $\mathbf{M}$ is scaled such that the product of the normalized $\mathbf{A}$ and $\mathbf{M}$ comes as close as possible to the original image $\mathbf{Y}$. This is achieved by minimizing

$$Q_\mathrm{s} = \|\mathbf{M}\,\mathrm{diag}(\mathbf{g})\mathbf{A} - \mathbf{Y}\|_\mathrm{F}^2 = \|\sum_{r=1}^{R} g_r \mathbf{M}_r \mathbf{A}_r - \mathbf{Y}\|_\mathrm{F}^2 \tag{4.60}$$

$$= \mathrm{tr}\left(\left(\sum_{r=1}^{R} g_r \mathbf{M}_r \mathbf{A}_r - \mathbf{Y}\right)\left(\sum_{r'=1}^{R} g_{r'} \mathbf{M}_{r'} \mathbf{A}_{r'} - \mathbf{Y}\right)^\mathrm{T}\right)$$

with $\mathbf{g} \in \mathbb{R}^R, \mathbf{g} \geq \mathbf{0}$. The elements of the vector $\mathbf{g}$ are the elements of the main diagonal of $\mathrm{diag}(\mathbf{g})$, all other elements are zero. The vectors $\mathbf{M}_r$ and $\mathbf{A}_r$ are the r-th row and column of $\mathbf{M}$ and $\mathbf{A}$, respectively. The scaling objective function (4.60) is quadratic, and due to the fact that it also is a composition of an affine mapping and a convex function, it is convex. For applying a gradient based minimization procedure (in this case, also PQN is used), the gradient is required. Its value with respect to one element of $\mathbf{g}$ is calculated by

$$\frac{\partial Q_\mathrm{s}}{\partial g_r} = 2 \sum_{r'=1}^{R} g_{r'} \,\mathrm{tr}\left(\mathbf{M}_r \mathbf{A}_r \mathbf{A}_{r'}^\mathrm{T} \mathbf{M}_{r'}^\mathrm{T}\right) - 2\,\mathrm{tr}\left(\mathbf{M}_r \mathbf{A}_r \mathbf{Y}^\mathrm{T}\right) . \tag{4.61}$$

The variable B (see Sec. 4.4.4) defines the width of the square column in the bottom graphic of Fig. 4.15 that is used for calculating the $\mathbf{N}_j$. A small value of B, for instance 5, will probably result in the consideration of high-SNR coefficients, while $B = \min\{X, Y\}$ considers also the coefficients with low SNR, but can lead to a better signal shape. Each value of B therefore leads to a slightly different matrix $\mathbf{M}$. The objective function (4.60) is solved for different values of B, all providing a different objective function value. The value of B that results in the lowest value and its corresponding matrix $\mathbf{M}$ is selected. This matrix and the normalized matrix $\mathbf{A}$ represent the final unmixing outcome of 3DDCT-NIMF.

4.4.6 Preprocessing and Full Algorithm

As with all other unmixing methods described in this thesis, it is reasonable to normalize the image matrix $\mathbf{Y}$ before applying 3DDCT-NIMF such that the spectra of each considered image have roughly the same magnitude, no matter if they represent relative reflectance or not. This makes it possible to constrain the range of possible regularization parameter values to the same for all considered images and enables a quick search for the most suited one. As 3DDCT-NIMF involves the calculation of the 3D DCT, the frequency coefficient of $f_x = f_y = f_l = 0$ is the bias and is very likely to have by far the highest absolute value of all coefficients. For avoiding improper scaling and numerical problems during optimization, the mean of the normalized image matrix $\mathbf{Y}$ is subtracted before the calculations and added to the final $\mathbf{M}$ again.

3DDCT-NIMF requires high accuracy in estimating the $\mathbf{N}_j$ for ensuring that $\mathbf{M}$ is an exact estimate after blockwise inversion and inverse DCT. Estimating a single element of $\mathbf{N}_j$ just slightly wrong could greatly falsify the inverted matrix. For this reason, the optimization is performed for a large number T of iterations, typically a few thousand. The matrices $\mathbf{A}$ and $\mathbf{N}_j$ are optimized alternatingly with 5 PQN steps each, i.e., five PQN steps for $\mathbf{A}$, and subsequently five steps for each $\mathbf{N}_j$. For increasing the accuracy of the final outcome, the number of PQN steps is altered to 25 for $\mathbf{A}$ and 100 for $\mathbf{N}_j$ during the last 10 alternating abundance/endmember iterations.

The full 3DDCT-NIMF procedure is given in Algorithm 2.

4.4.7 Further Regularizer Convexification

The regularizer $-\log(|\det(\mathbf{N}_j)|)$ in (4.41) is not convex. As stated before, it is derived from maximum likelihood considerations, and mainly arises from the inverse transform, i.e., the matrix inversion. Apart from representing the inverse transform, its goal is to prevent the matrices $\mathbf{A}$ and $\mathbf{N}_j$ from turning $\mathbf{0}$, see Sec. 4.4.3. As this regularizer is not convex, the objective function (4.41) only has a convex data fidelity part, but is not convex in total. The following considerations aim at deriving a convex replacement, which, if it is not possible to take the stochastic properties of the inverse transform fully into account, at least ensures that the $\mathbf{N}_j$ are invertible. A thorough investigation of this approach, however, goes beyond the scope of the thesis

It turns out to be a quite difficult task to ensure matrix invertibility in a convex manner during optimization. Instead of considering the determinant by, e.g., including $\det(\mathbf{N}_j) > 0$ as a constraint, it is promising to consider the singular values and eigenvalues of a matrix. In the following, the index j will be dropped, and it will be assumed that $\mathbf{N}$ is an arbitrary complex matrix $\mathbf{N} \in \mathbb{C}^{R \times R}$. Its eigenvalues are denoted by $\lambda_1, \ldots, \lambda_R$ and sorted in descending order: $|\lambda_1| \geq \cdots \geq |\lambda_R|$. Correspondingly, its singular values are denoted by $\sigma_1, \ldots, \sigma_R$ and sorted: $\sigma_1 \geq \cdots \geq \sigma_R$. Using

Algorithm 2 3DDCT-NIMF procedure.

1) Input: Hyperspectral image matrix $\mathbf{Y}$, initial endmember matrix $\mathbf{M}_{\text{init}}$, initial abundance matrix $\mathbf{A}_{\text{init}} = 1/R$ (same value for all elements)
2) Normalize $\mathbf{Y}$ such that its mean equals 1
3) Subtract the bias from $\mathbf{Y}$
4) Calculate $\mathbf{Y}_{\text{3DDCT}}$, the 3D DCT of $\mathbf{Y}$
5) Calculate the absolute values of all 3D DCT coefficients and retain the ratio ρ of the largest coefficients; set all others to zero.
6) Split $\mathbf{Y}_{\text{3DDCT}}$ into J blocks $\mathbf{Y}_j$
7) Given an initial endmember matrix $\mathbf{M}_{\text{init}}$, calculate the $\mathbf{N}_j$ by performing 1D DCT column by column, splitting the resulting matrix into J blocks and inverting them
8) **For** $t = 1$ **to** $T - 10$ **do**
   Minimize (4.41) in an alternating fashion
      Perform 5 PQN steps optimizing $\mathbf{A}$
      Perform 5 PQN steps optimizing each $\mathbf{N}_j$
   end for
9) Decrease stopping criteria for higher accuracy
10) **For** $t = T - 9$ **to** T **do**
   Minimize (4.41) in an alternating fashion
      Perform 25 PQN steps minimizing $\mathbf{A}$
      Perform 100 PQN steps minimizing each $\mathbf{N}_j$
   end for
11) Calculate the final matrix $\mathbf{A}$ from the first block, i.e., minimize (4.41) for $j = 1$
12) **For** $B = 4, 6, 8, 10, 15, 20, \ldots \min\{X, Y\}$ **do**
   Recalculate the $\mathbf{N}_j$ using (4.41) with adapted $\mathbf{W}_{XY,j}$
   Create a new $L \times R$ matrix by stacking the inverted
      matrices $\mathbf{N}_j^{-1}$ on top of each other
   Perform columnwise 1D IDCT for calculating $\mathbf{M}$
   Subtract the mean of $\mathbf{M}$, as $\mathbf{Y}$ also has zero mean
   Minimize (4.60), define scaled $\mathbf{M} := \mathbf{M}\,\text{diag}(\mathbf{g})$
   Add mean of normalized original image to $\mathbf{M}$
   end for
13) Select the B that provides the lowest value of (4.60)
14) Output: calculated matrices $\mathbf{M}$, $\mathbf{A}$

Weyl's inequalities [139, 149],

$$\prod_{i=1}^{I} |\lambda_i| \leq \prod_{i=1}^{I} \sigma_i, \quad I = 1, 2, \dots R-1, \tag{4.62}$$

$$\prod_{i=1}^{R} |\lambda_i| = \prod_{i=1}^{R} \sigma_i, \tag{4.63}$$

a relation between the smallest singular value and the smallest eigenvalue can be constructed. Let $a := \prod_{i=1}^{R-1} |\lambda_i|$, $b := \prod_{i=1}^{R-1} \sigma_i$. Using (4.62) yields $a \leq b$. From (4.63) follows $a\,|\lambda_r| = b\sigma_R$, and thus

$$|\lambda_R| \geq \sigma_R. \tag{4.64}$$

By this equation, an inequality between the smallest singular value and the smallest eigenvalue is established. This means that if the smallest singular value is forced to be > 0, the smallest eigenvalue will also be > 0. Considering that the matrix determinant equals the product of its eigenvalues,

$$\det(\mathbf{N}) = \lambda_1 \lambda_2 \cdots \lambda_R, \tag{4.65}$$

it becomes clear that ensuring that the smallest singular value $\sigma_R > 0$ ensures that the matrix is invertible.

One option for ensuring $\sigma_R \geq 0$ is given in [128]. In the following, let $\|\mathbf{N}\|_2 = \sigma_1$ denote the spectral norm, i.e., the largest singular value of $\mathbf{N}$. Interestingly, it holds $\|\mathbf{N}^{-1}\|_2 = \dfrac{1}{\sigma_R}$, i.e., the spectral norm of an inverse matrix is the reciprocal value of the smallest singular value of the original matrix. Let us now consider the given matrices $\mathbf{R}, \mathbf{N} \in \mathbb{C}^{R \times R}$. According to [128], supposing

$$\|\mathbf{I} - \mathbf{R}\mathbf{N}\|_2 \leq \alpha < 1, \tag{4.66}$$

the following inequalities hold:

$$\frac{\|\mathbf{R}\|_2}{1+\alpha} \leq \|\mathbf{N}^{-1}\|_2 \leq \frac{\|\mathbf{R}\|_2}{1-\alpha}. \tag{4.67}$$

Using the spectral norm of the inverse matrix yields

$$\frac{1}{\sigma_R} \leq \frac{\|\mathbf{R}\|_2}{1-\alpha} \Rightarrow \sigma_R \geq \frac{1-\alpha}{\|\mathbf{R}\|_2} . \tag{4.68}$$

This means that it is ensured that the smallest singular value of $\mathbf{N}$, σ_R, is larger than or equal to a positive number, as $\mathbf{R}$ can be chosen arbitrarily. As a consequence, $\mathbf{N}$ is regular, i.e., invertible. This requires the incorporation of the constraint (4.66) into the optimization, which is required for (4.67) to hold. One option is to choose $\mathbf{R}$ to be the identity matrix, yielding $\|\mathbf{R}\|_2 = 1$ and simplifying the constraint to $\|\mathbf{I} - \mathbf{N}\|_2 \leq \alpha$ ($\alpha < 1$). It is reasonable to choose α to a value close to 1, such as 0.99, to allow for maximum flexibility of $\mathbf{N}$.

According to [19] (see also Sec. 2.2.4), an optimization problem is convex if the objective function is convex [19, Sec. 4.2.1] and the inequality constraints are of the form $f_i(\mathbf{x}) \leq 0$, where the $f_i(\mathbf{x})$ are convex functions. The constraint (4.66) can be reformulated to

$$\|\mathbf{I} - \mathbf{N}\|_2 - \alpha \leq 0 . \tag{4.69}$$

As all matrix norms are convex (see Sec. 4.4.3 and [19]), and $\mathbf{I}-\mathbf{N}$ represents an affine mapping, the function on the left side of (4.69) is convex. The full optimization problem then amounts to

$$\begin{aligned} &\text{minimize} \sum_{j=1}^{J} \|\mathbf{N}_j \mathbf{Y}_j - \mathbf{A}\mathbf{W}_{XY,j}\|_{\mathrm{F}}^2 \\ &\text{subject to } \mathbf{A} \geq \mathbf{0} , \\ &\text{subject to } \|\mathbf{I} - \mathbf{N}_j\|_2 - \alpha \leq 0, \ j = 1, \ldots, J . \end{aligned} \tag{4.70}$$

While optimizing this problem results in invertible matrices $\mathbf{N}_j$, the problem is that the constraint (4.69) allows these matrices to move only a certain distance away from the identity matrix; most likely, its main diagonal will have positive values and the other elements will be quite small.

As this is potentially limiting, another approach would be to constrain the $\mathbf{N}_j$ to be diagonally dominant. Diagonally dominant matrices have the property that the absolute value of each diagonal entry is larger than or equals the sum of the absolute values of all other elements in the respective row. Such matrices are invertible [57, Ex. 36 (c)], but the difficulty with this approach is that the constraint of having dominant diagonals has to be incorporated in a convex way, meaning that the diagonal elements have to be constrained to be either positive or negative (the sum of the absolute values of the other elements is convex). This requires to try all combinations of positive/negative diagonal elements for minimizing/maximizing them and decide for the best combination.

The final, proposed approach is quite similar to diagonally dominant matrices, but does not impose any constraints on the non-diagonal elements. However, it also requires the trial and error of all positive/negative diagonal combinations.

According to [119], the function

$$f_\kappa(\mathbf{N}) = \sum_{i=1}^{\kappa} \sigma_i \tag{4.71}$$

is the sum of the κ largest singular values and

$$f_R(\mathbf{N}) = \sum_{i=1}^{R} \sigma_i \tag{4.72}$$

the sum of all singular values. Using these two equations, the smallest singular value is given by

$$\sigma_R = f_R(\mathbf{N}) - f_{R-1}(\mathbf{N}) = \sum_{i=1}^{R} \sigma_i - \sum_{i=1}^{R-1} \sigma_i \,. \tag{4.73}$$

The sum $f_\kappa(\mathbf{N})$ is convex for all $\kappa = 1 \ldots R$, see the proof in [119]. For this reason, (4.73) is a convex function ($\sum_{i=1}^{R} \sigma_i$) plus a concave one ($-\sum_{i=1}^{R-1} \sigma_i$). As the smallest singular value (4.73) is to be maximized, or, more precisely, should be larger than 0, it should be a concave function.

Also given in [119] is the relation $\operatorname{atr}(\mathbf{N}) \leq f_R(\mathbf{N})$, where $\operatorname{atr}(\mathbf{N})$ denotes the absolute trace, i.e., the sum of the absolute diagonal elements: $\operatorname{atr}(\mathbf{N}) = \sum_{r=1}^{R} |N_{rr}|$. Therefore, it is possible to replace the convex part by a linear part, namely, the matrix trace $\operatorname{tr}(\mathbf{N}) = \sum_{r=1}^{R} N_{rr}$. The common matrix trace represents a lower bound of the absolute trace, because the absolute trace will always be larger than or equal to the value of the common trace. As the trace is a linear function of the (diagonal) elements of $\mathbf{N}$, it is both convex and concave at the same time. The function that is to be minimized (i.e., its negative is maximized) in total amounts to

$$Q_{\text{reg}} = -\left(\text{tr}(\mathbf{N} \odot \mathbf{S}) - \sum_{i=1}^{R-1} \sigma_i\right) . \tag{4.74}$$

Matrix $\mathbf{S}$ is a diagonal matrix with diagonal elements ± 1. By changing the signs of these diagonal elements (++++, +++-, ++-+ etc.), it is possible to either minimize or maximize the diagonal elements without any requirements on the non-diagonal elements. In turn, this maximizes the absolute trace and therefore the minimum singular value. The full objective function is given by

$$Q_{\text{3DDCT}} = \sum_{j=1}^{J} \|\mathbf{N}_j \mathbf{Y}_j - \mathbf{A}\mathbf{W}_{XY,j}\|_{\text{F}}^2 - K_j \left(\text{tr}(\mathbf{N}_j \odot \mathbf{S}) - \sum_{i=1}^{R-1} \sigma_i\right) . \tag{4.75}$$

In contrast to 3DDCT-NIMF as presented in Sections 4.4.3–4.4.6, the sequence of abundance and endmember estimation is slightly changed. The abundance matrix $\mathbf{A}$ and the matrices $\mathbf{N}_j$ are calculated using (4.75). For each matrix $\mathbf{N}_j$, the optimal signs of each diagonal element are determined by trying all combinations of positive/negative sign for a few optimization iterations, maybe 10, and the combination providing the lowest value of the data fidelity term $\|\mathbf{N}_j \mathbf{Y}_j - \mathbf{A}\mathbf{W}_{XY,j}\|_{\text{F}}^2$ in (4.75) is selected. For each combination run, all diagonal elements of the $\mathbf{N}_j$ are initially set to zero for fairness between positive and negative values. Afterwards, the best sign combinations of each block j are used to calculate the regular 3DDCT-NIMF procedure as described in the previous section for determining $\mathbf{A}$ and $\mathbf{M}$, but this time using the objective function (4.75).

Alternatively, the abundance matrix $\mathbf{A}$ and the first matrix $\mathbf{N}_1$ can be calculated from the first block, searching for the optimal sign combination of $\mathbf{N}_1$ with a few iterations first and performing full optimization for $\mathbf{A}$ and $\mathbf{N}_1$ afterwards. The resulting matrix $\mathbf{A}$ can be fixed, and the best sign combinations for the matrices $\mathbf{N}_2$, $\mathbf{N}_3$ etc. determined with this fixed $\mathbf{A}$. The advantage is that the blocks are decoupled, meaning that a fair search for the best signs that is not influenced by the other blocks is performed.

Maybe this is also a disadvantage of this approach, as the matrix $\mathbf{A}$ is calculated without the full information contained in all J blocks.

For completeness, the gradient of the regularizer of (4.75), i.e., (4.74) is given by

$$\frac{\partial Q_{\text{reg}}}{\partial \mathbf{N}_j} = -(\mathbf{I} \odot \mathbf{S}) + \mathbf{U}_{j,(:,1:R-1)} \cdot \left(\mathbf{V}_{j,(:,1:R-1)}\right)^{\mathrm{T}}. \tag{4.76}$$

The matrices $\mathbf{U}_j$ and $\mathbf{V}_j$ result from the singular value decomposition $\mathbf{N}_j = \mathbf{U}_j \mathbf{\Sigma}_j \mathbf{V}_j^{\mathrm{T}}$ [108]; only the first $R-1$ columns are used.

The goal of this section was to present some thoughts and a method how the first 3DDCT-NIMF procedure described in Sections 4.4.3–4.4.6 can be converted to a fully convex optimization problem. It goes beyond the scope of this thesis to fully analyze the discussed approach. The original 3DDCT-NIMF and its results will be analyzed in Sec. 5.6. Section 6.2 will discuss how the new approach could be incorporated for future methods.

4.5 3D Singular Value Decomposition (SVD) Unmixing

In Sec. 4.4.4, the characteristics of the noise of the 3D DCT transformed image have been analyzed. In case the noise added to all image elements is uncorrelated and normally distributed with the same variance, it was deduced that the noise of the transformed image elements follows the same distribution. Interestingly, this consideration does not only hold for the DCT, but also for all other transforms that can be expressed by multiplication of the data matrix with an orthogonal transformation matrix, see equations (4.56) to (4.59). Out of the many possible orthogonal transforms, the singular value decomposition (SVD) is also analyzed in this thesis. Other possible transforms include, e.g., the wavelet transform and principal component analysis (PCA). Note, however, that PCA cannot easily be applied to all three image dimensions, because it is necessary to subtract the mean first. While this is not an issue for the spectral dimension (the subtracted mean can be added to the estimated endmembers after the unmixing), it is not straightforward how to incorporate a spatial mean that has been calculated from a spectrally transformed image tensor. For using the SVD, the mean does not need to be considered separately.

By SVD, every matrix $\mathbf{Y} \in \mathbb{R}^{L \times I}$ can be decomposed into the product

$$\mathbf{Y} = \mathbf{U}\boldsymbol{\Sigma}\mathbf{V}^{\mathrm{T}} . \tag{4.77}$$

The matrices $\mathbf{U} \in \mathbb{R}^{L \times L}$ and $\mathbf{V} \in \mathbb{R}^{I \times I}$ are orthogonal matrices. The diagonal of matrix $\boldsymbol{\Sigma} \in \mathbb{R}^{L \times I}$ are the singular values of $\mathbf{Y}$ in decreasing order, while all other elements equal zero. It is straightforward to multiply the image matrix $\mathbf{Y}$ with the matrix $\mathbf{U}^{\mathrm{T}}$, and from the fact that $\mathbf{U}$ is orthogonal it follows

$$\mathbf{U}^{\mathrm{T}}\mathbf{Y} = \mathbf{U}^{\mathrm{T}}\mathbf{U}\boldsymbol{\Sigma}\mathbf{V}^{\mathrm{T}} = \boldsymbol{\Sigma}\mathbf{V}^{\mathrm{T}} , \tag{4.78}$$

so the matrix $\mathbf{U}$ obtained by singular value decomposition of $\mathbf{Y}$ is used as orthogonal transformation matrix.

By analogy with the 3D discrete cosine transform, a 3D SVD transform can be applied to the hyperspectral image tensor $\mathcal{Y}$ first; the subsequent unmixing procedure is the same as described in the previous section. In case of the 3D SVD, the first SVD is calculated along the spectral axis, i.e., from the unfolded matrix $\mathbf{Y}$. This is due to the fact that all pixel spectra are superpositions of the pure endmember spectra and therefore exhibit a large degree of redundancy. The spectral redundancy enables significant data reduction; by contrast to the 3D DCT, it is not necessary to consider the full 3D dimensionality. While 3D discrete cosine transform nonnegative inverse matrix factorization (3DDCT-NIMF) requires the consideration of the full column in Fig. 4.15 for accurately calculating the endmembers, the SVD compresses all spectral details into the first few blocks. For this reason, it is now sufficient to set the number J of blocks considered in the objective function (4.41) to a low value of 1–5. All other blocks are likely to consist of very small values that contain a large ratio of noise energy and can therefore be disregarded. After applying the first SVD along the spectral axis, the two spatial SVDs are calculated subsequently along both spatial axes of the resulting image tensor. The only difference between this procedure and 3DDCT-NIMF is the different choice of the transformation matrices; all other steps, for instance the scaling ambiguity resolution (Sec. 4.4.5), remain the same. The new SVD unmixing method is termed 3DSVD-NIMF.

4.6 Objective Function Minimization

This section gives a short overview on some minimization algorithms that have already been used in spectral unmixing. In addition, it proposes the use of one that, to the best of the author's knowledge, has not been used for unmixing before.

4.6.1 Minimization Techniques

After having introduced various unmixing objective functions in the previous sections, it will now be analyzed how they can be minimized efficiently. For the following, MVC-NMF [95] described in Sec. 3.2.2 will be considered as spectral regularizer and combined with the HS spatial regularization from Sec. 4.1. This leads to the objective function

$$Q = \|\mathbf{MA} - \mathbf{Y}\|_{\mathrm{F}}^2 + \mu_{\mathrm{spec}} \frac{1}{2(R-1)!} \det{}^2 \left(\begin{bmatrix} \mathbf{1}_R^{\mathrm{T}} \\ \tilde{\mathbf{M}} \end{bmatrix} \right) + \mu_{\mathrm{spat}} \sum_{r=1}^{R} \sum_{x=1}^{X} \sum_{y=1}^{Y} \|H(\mathbf{A}_r)(x,y)\|_{\mathrm{F}}^2 . \tag{4.79}$$

The most widely used approach to minimizing the general objective function (3.2) consists of first fixing $\mathbf{M}$ and minimizing for $\mathbf{A}$ and vice versa. Such a sequence of alternating steps is continued until convergence is achieved. Due to the fact that (3.2) is not convex for both $\mathbf{M}$ and $\mathbf{A}$ simultaneously, there might be more than one minimum and depending on the chosen minimization method and the chosen optimization scheme (e.g., there can be multiple minimization steps for $\mathbf{M}$ before $\mathbf{A}$ is updated again), the used algorithms can come to different solutions.

Many NMF algorithms [53, 61, 83] are based on gradient methods, i.e., the minimization is performed iteratively by taking steps into the direction of the negative gradient. The update of $\mathbf{A}$ is calculated by

$$\mathbf{A}^{(p+1)} = \mathbf{A}^{(p)} - \delta \frac{\partial Q}{\partial \mathbf{A}} , \tag{4.80}$$

where δ is the step size and p denotes the iteration number. The update of $\mathbf{M}$ is calculated analogously.

Possible choices for the step size δ are a constant step size or a diminishing step size $\delta = k/p$, where k is a constant. The HS implementation in [176] (see also Section 4.1) uses a diminishing step size. On the one hand, such rules have the advantage that they perform fast as no step size calculation is required. On the other hand, they have the disadvantage that the constant k should be chosen feasibly; otherwise, many steps are required for the precise calculation of the minimum that is demanded by NMF, see Fig. 4.16. An adaptive step size rule with guaranteed objective reduction is the Armijo rule [10, 11]. In their respective original publications, the spectrally regularized methods MVC-NMF [95], MiniDisCo-NMF [63] and MDC-NMF [154] minimize using this rule.

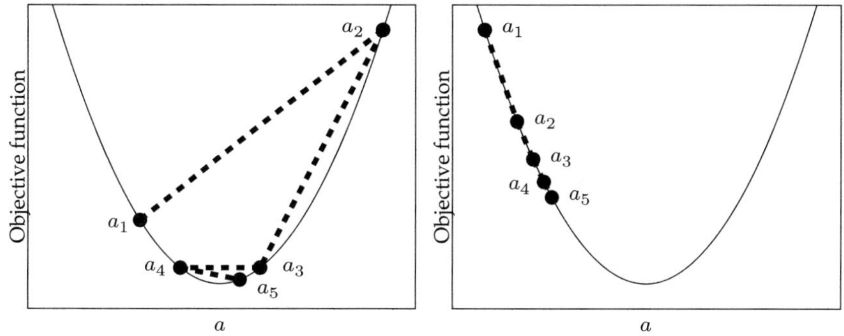

Figure 4.16 One-dimensional objective function minimization with diminishing step size. Too large initial step size (left), too small initial step size (right). Although the left graph looks more or less acceptable, the minimum is not found very accurately, as the method jumps between both arms of the function and does not converge quickly. This problem becomes more evident when the initial step size is chosen even larger.

Due to the fact that minimization methods are most often described with the sought-after parameters arranged to a vector, this notation will be used for the explanation of the Armijo rule. Note that the matrix elements of $\mathbf{M}$ and $\mathbf{A}$ can be rearranged to vectors; the minimization principle is the same.

Before taking the minimization step, the Armijo rule checks if the initial step size leads to a sufficient decrease of the objective function. If this is the case, the step size is increased. If this is not the case, the step size is reduced by the factor β^m, where $0 < \beta < 1$ and m is the number of attempts in the search for a feasible step size. For an arbitrary continuously differentiable objective function $Q(\mathbf{a})$, the condition for the sufficient decrease is

$$Q(\mathbf{a}^{(p)}) - Q(\mathbf{a}^{(p)}(s\beta^m)) \geq \sigma \nabla Q(\mathbf{a}^{(p)})^{\mathrm{T}} \left(\mathbf{a}^{(p)} - \mathbf{a}^{(p)}(s\beta^m)\right) \quad (4.81)$$

with the point $\mathbf{a}^{(p)} \in \mathbb{R}^n$ at iteration p. Besides, (4.81) contains the initial step size s, the sufficient decrease parameter $\sigma > 0$ and the objective function gradient $\nabla Q(\mathbf{a})$ at position $\mathbf{a}$. For $m = 1, 2, \ldots$, it is tested if condition (4.81) is fulfilled; if not, m is increased by one etc. Setting $\delta^{(p)} = s\beta^m$, the update $\mathbf{a}^{(p+1)}$ is calculated by

$$\mathbf{a}^{(p+1)} = \mathbf{a}^{(p)}(\delta^{(p)}) = \left(\mathbf{a}^{(p)} - \delta^{(p)} \nabla Q(\mathbf{a})\right)_{\mathrm{P}}, \quad (4.82)$$

where $(\cdot)_{\mathrm{P}}$ denotes the projection onto a closed convex subset of $\mathbb{R}^n$. This projection is required for enforcing the NN and STO constraints on $\mathbf{M}$ and $\mathbf{A}$ from Section 4. In this specific case, the subset is the probability simplex.

In general, it is also possible to choose the step size δ such that the update (4.80) is no longer an additive update, but becomes a multiplicative one. In this case, the matrix $\mathbf{A}^{(p)}$ from the last step and the term $-\delta \frac{\partial Q}{\partial \mathbf{A}}$ can be merged and simplified to a product of matrix elements. This kind of update was proposed by Lee and Seung [77] and is also often used for unmixing methods employing NMF. One drawback of the multiplicative update is that when regularizers $\mathcal{R}_{\mathrm{spec}}$ or $\mathcal{R}_{\mathrm{spat}}$ are present, it is not always straightforward to incorporate them, as $\mathbf{A}^{(p)}$ and $-\delta \frac{\partial Q}{\partial \mathbf{A}}$ cannot be merged that easily because the derivative may have additional terms not straightforwardly expressable in dependence of the matrix elements. One example is the matrix determinant. Due to the fact that this thesis mainly deals with new regularizers, this kind of update will not be used.

A minimization method allowing for easy implementation and providing efficient calculation is the limited-memory projected quasi-Newton method (PQN) [129]. Apart from the comparison in the following section, it is employed for both endmember and abundance updates throughout this work. It uses a quadratic model of the objective function that is minimized using a positive definite approximation of the Hessian of the objective function. There are two drawbacks of approximating the original objective function by a quadratic model and minimizing the quadratic model. First, the minimization of the model can be almost as difficult to solve as the original objective, and second, the size of the Hessian and therefore the size of its approximation grows quadratically with the number of considered parameters. For dealing with the second aspect, the approximation of the Hessian is calculated by the limited-memory variant of the Broyden-Fletcher-Goldfarb-Shanno (BFGS)-method [105]. Concerning the first aspect, a spectral projected gradient (SPG) method is used for the optimization of the quadratic model. The sufficient descent is determined based on several past iterations and not only on the last one, which allows for the objective function to increase in some steps, while general convergence is still guaranteed.

4.6.2 Comparison

Both the Armijo rule and the PQN method are applied to the minimization of the objective function (4.79); the method denoted by Armijo in the following minimizes both $\mathbf{M}$ and $\mathbf{A}$ using the Armijo rule, while the method called PQN employs PQN for both endmember and abundance updates. The *MATLAB* PQN implementation provided by [118] is used.

Both procedures share the same sequence: First, $\mathbf{A}$ is updated until the relative change of the objective function from one iteration to the next is below the threshold $J_{\mathrm{relchange}}$. After this, $\mathbf{M}$ is updated until the aforementioned condition is met. This procedure goes on until the objective value at the end of one series of optimization steps for either $\mathbf{M}$ or $\mathbf{A}$ changed less than $J_{\mathrm{relchange}}$ in comparison with the final objective function value of the previous $\mathbf{M}$ or $\mathbf{A}$ series. Note that there has been an estimation series of the respective other matrix in between that probably has increased the objective function value during the jump from $\mathbf{M}$ update to $\mathbf{A}$ update or vice versa, so the final value of the $\mathbf{M}$ update has to be compared with the final value of the last $\mathbf{M}$ update and not of the last $\mathbf{A}$ update. When this condition on two subsequent $\mathbf{M}$ or $\mathbf{A}$ series is met, the whole optimization finishes. The maximum number of iterations for $\mathbf{M}$ or $\mathbf{A}$ is 100, while the maximum number of $\mathbf{M}/\mathbf{A}$ sequences is 200 for Armijo and 600 for PQN. In many cases, however, the maximum iteration numbers are not reached, because the stopping criterion based on the relative objective value is fulfilled before. The value of $J_{\mathrm{relchange}}$ is set to 10^{-8}.

Both the Armijo and the PQN methods are applied to the Jasper and the Urban hyperspectral images described in Sec. 5.1.1. The abundance root-mean-square error (ABU RMSE) and endmember spectral angle (SA) (see Section 5.2) will be used as unmixing quality measures.

Figure 4.17 shows the comparison between both methods applied to the Jasper image degraded by bandwise noise of 30 dB SNR. It can be seen that the obtained endmember SA and ABU RMSE values are quite similar. PQN achieves for almost all considered combinations of μ_{spec} and μ_{spat} lower final objective function values than Armijo. Note that when the ground truth is not known, the objective function is the only measure that lets one compare the outcome of different methods. PQN needs significantly less calculation time than Armijo; for all considered parameter configurations, the average time saving of PQN is about 20 %. Close to the optimal values of μ_{spec} and μ_{spat}, the saving is more than 50 %. From the figure, it can be seen that the spectral regularization of MVC-NMF has a larger impact than the spatial regularization using HS. The optimal values for ABU RMSE and endmember SA are obtained for $\mu_{\text{spat}} = 0.002$ and $\mu_{\text{spec}} = 0.5$ for PQN (RMSE = 0.0277) and $\mu_{\text{spec}} = 1$ in case the Armijo rule is applied (RMSE = 0.0346). The objective function values during minimization ($\mu_{\text{spat}} = 0.002$, $\mu_{\text{spec}} = 0.5$) are displayed in Fig. 4.18. It can be seen that the PQN minimization works faster and obtains a lower final objective function value. Note that the stopping criteria for series of pure $\mathbf{M}$ or $\mathbf{A}$ minimizations are the same for both methods to allow for a fair comparison; only the number of $\mathbf{M}/\mathbf{A}$ sequences differs.

The Urban image has more pixels than the Jasper image (307×307 versus 100×100). $J_{\text{relchange}}$ of PQN was lowered to 10^{-9} because of the flatter objective function shape close to the minimum. The unmixing results in dependence of μ_{spec} and μ_{spat} (SNR 10 dB) show a similar shape as for the Jasper image. Due to the high noise level, the HS regularization becomes more important; the best RMSE value is at $\mu_{\text{spat}} = 0.01$ (the optimal μ_{spec} changed to 2000). The best SA value, however, is obtained for $\mu_{\text{spat}} = 0$. For this image, the time saving of PQN in comparison with Armijo close to the optimal values of μ_{spec} and μ_{spat} is only about 10 %. Figure 4.19 shows the time difference between the two methods. Close to the minimum, the convergence speed and accuracy of PQN decreases. This could be due to the objective function approximation of PQN that becomes inaccurate

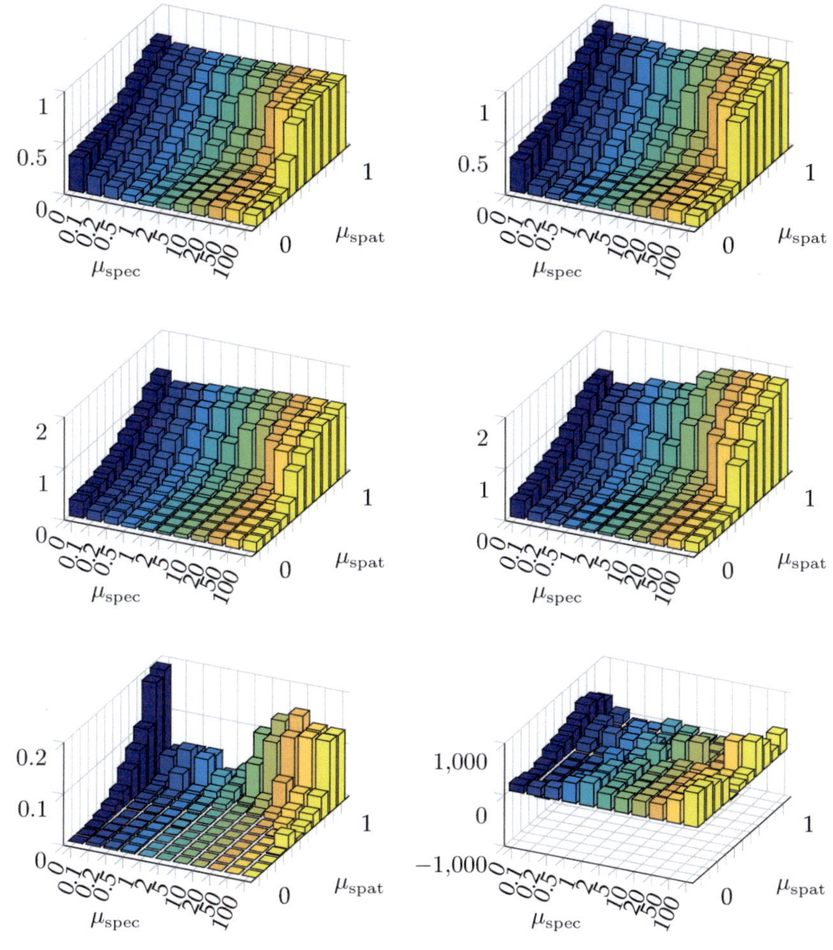

Figure 4.17 Jasper image at 30 dB noise level: Comparison between Armijo and PQN. First row: Armijo ABU RMSE (left), PQN ABU RMSE (right). Second row: Armijo endmember spectral angle (left), PQN endmember spectral angle (right). Third row: relative final objective function value ((Armijo-PQN)/PQN) (left), time difference in seconds (calculation time Armijo-calculation time PQN) (right). The tick labels of μ_{spat} between 0 and 1 are omitted for better readability. They are 0.001, 0.002, 0.005, 0.01, 0.02, 0.05, 0.1, 0.2 and 0.5.

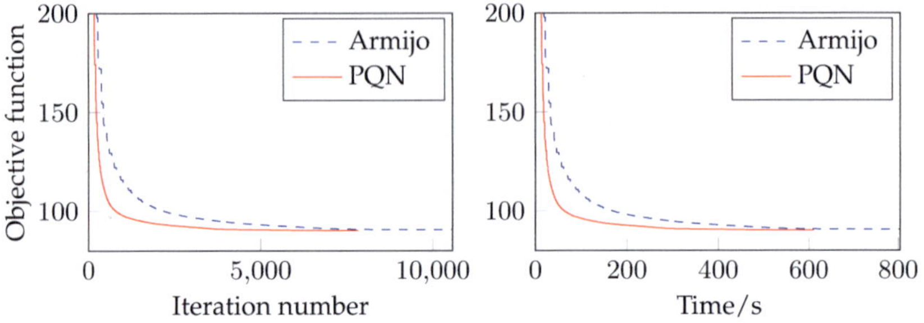

Figure 4.18 Jasper image at 30 dB noise level: Objective function over iteration number and time. Note the faster convergence and lower final value of the PQN minimization in both cases.

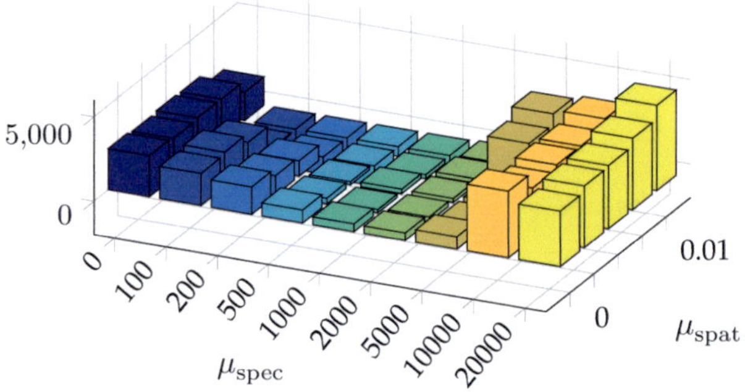

Figure 4.19 Urban image at 10 dB noise level: Time difference between Armijo and PQN in seconds. The omitted tick labels of $\mu_{spat} = 0.002$ are 0.001, 0.002 and 0.005.

at high noise levels. From these short analyses, it can be concluded that PQN leads to a significant reduction of the calculation time, while mostly obtaining final objective function values very similar to the minimization based on the Armijo rule and even lower ones in the case of low noise levels. This justifies the use of PQN for the unmixing purposes described in this thesis.

4.7 Multidimensional Image Denoising by Weighted Addition of Images (MIDWAI)

This section describes a multidimensional image denoising algorithm developed during a master's thesis [194]. It is based on the weighted addition of 2D subimages, therefore termed multidimensional image denoising by weighted addition of images (MIDWAI). In [194], only a version designed for removing Poisson noise is given whose results have been published in [164]. Here, an unpublished extension dealing with the presence of Gaussian noise is given. Additionally, denoising results considering not only 3D hyperspectral images, but also four-dimensional hyperspectral fluorescence images are presented. The fourth dimension of the fluorescence images is the used illumination wavelength. The proposed method can be applied to multidimensional images other than conventional hyperspectral or hyperspectral fluorescence images.

Similar to the method described in [76], the core contribution of MIDWAI is to not extend powerful 2D denoising techniques to 3D, but to adequately project the additional dimensions on 2D images. In case of hyperspectral images, this is based on the observation that the spectral dimension exhibits less variability than the spatial [76, 106, 148]. If the denoising strength of the projection should not be sufficient, additional 2D techniques can be applied afterwards. Another possibility is to use the 3D methods described in [30, 106] that require images with low noise. With the presented method, these low-noise images can be calculated very efficiently.

Depending on the chosen parameter values, MIDWAI does not necessarily take all the information that is present in a multidimensional image into account. In many cases, however, this is not required, as the already used redundancy is sufficient and further information cannot be exploited easily or leads to no further improvement.

The major random noise sources in digital image acquisition are [71, 89]

- the stochastic nature of the photon arrivals at each pixel and
- thermal and electronic fluctuations within the acquisition devices.

The photon arrivals at each pixel can be modelled by a Poisson random process whose rate μ is determined by the light intensity [152], while the noise added during acquisition is assumed to be additive white Gaussian noise (AWGN) [89]. In contrast to conventional image sensors, the images obtained with EMCCD sensors show Poisson characteristics for both high and low signal intensity, see Sec. 2.3.

As the image noise can follow either a Gaussian or a Poisson characteristic, two versions of the MIDWAI procedure are introduced, one that can deal with Poisson noise, and another, more general one that is well suited for Gaussian noise and other noise types that are sufficiently well characterized by their mean and variance.

4.7.1 MIDWAI for Poisson Noise

Without loss of generality, three-dimensional hyperspectral images having two spatial and one spectral dimensions are considered exemplarily. Extensions to images of higher dimensionality are straightforward.

The hyperspectral image that is to be denoised is denoted by $\mathcal{Y} \in \mathbb{R}^{X\times Y\times L}$. It consists of L two-dimensional subimages $\mathbf{Y}_l \in \mathbb{R}^{X\times Y}$ $(1 \leq l \leq L)$, where L denotes the number of wavelengths and X and Y denote the total number of pixels in the spatial directions. In the following, matrices of random variables and constants will be calculated and manipulated. This means that all operations such as addition, subtraction, multiplication, division, squaring, calculating the expected value and the variance, etc., are performed elementwise. For clarity of notation, standard notation of these operations will be used. It will be clear from the context whenever elementwise operations are to be used.

The subimage that is to be denoised presently is $\mathbf{Y}_l$. The other ones can be treated the same way afterwards. Let us now assume that the denoising of $\mathbf{Y}_l$ is performed by adding the image $\mathbf{Y}_{l+1}$. The elements of $\mathbf{Y}_l$ and $\mathbf{Y}_{l+1}$ depict realizations of the Poisson distributed random variables $\mathtt{Y}_l$ and $\mathtt{Y}_{l+1}$ having the means $\mathbf{\Lambda}_l$ and $\mathbf{\Lambda}_{l+1}$. All matrices $\mathbf{\Lambda}_l$, $\mathbf{\Lambda}_{l+1}$, $\mathtt{Y}_l$ and $\mathtt{Y}_{l+1}$, etc., are of the same size as $\mathbf{Y}_l$.

The objective is to find an estimator $\widehat{\mathbf{\Lambda}}_l$ for the matrix of true means $\mathbf{\Lambda}_l$. As both ideal (i.e., the images that are sought after by the denoising process) two-dimensional images at bands k and $k+1$ in general do not have the same image values at the same pixel position, the idea is to weightedly add both images. The ansatz is

$$\widehat{\mathbf{\Lambda}}_l = \frac{\mathrm{Y}_l + \mathbf{B}_{l+1}\mathbf{C}_{l+1}\mathrm{Y}_{l+1}}{1+\mathbf{B}_{l+1}} \tag{4.83}$$

with the spatially varying positive constants $\mathbf{B}_{l+1}$ and $\mathbf{C}_{l+1}$. One requirement is that the estimator has to provide an estimation of the true value, i.e., it is unbiased. From this requirement follows

$$\mathrm{E}\{\widehat{\mathbf{\Lambda}}_l\} = \frac{\mathbf{\Lambda}_l + \mathbf{B}_{l+1}\mathbf{C}_{l+1}\mathbf{\Lambda}_{l+1}}{1+\mathbf{B}_{l+1}} \stackrel{!}{=} \mathbf{\Lambda}_l \tag{4.84}$$

and therefore

$$\mathbf{C}_{l+1} = \frac{\mathbf{\Lambda}_l}{\mathbf{\Lambda}_{l+1}} \,. \tag{4.85}$$

Due to the unbiasedness of the estimator $\mathrm{E}\{\widehat{\mathbf{\Lambda}}_l\}$, its mean squared error (MSE) equals its variance [126]. For this reason, the variance of the estimator is minimized for minimizing the MSE:

$$\min_{\mathbf{B}_{l+1}} \mathrm{Var}\{\widehat{\mathbf{\Lambda}}_l\} = \min_{\mathbf{B}_{l+1}} \frac{\mathbf{\Lambda}_l + \mathbf{B}_{l+1}^2 \left(\frac{\mathbf{\Lambda}_l}{\mathbf{\Lambda}_{l+1}}\right)^2 \mathbf{\Lambda}_{l+1}}{\left(1+\mathbf{B}_{l+1}\right)^2} \,. \tag{4.86}$$

Deriving the variance in (4.86) with respect to $\mathbf{B}_{l+1}$ leads to

$$\frac{\partial \mathrm{Var}\{\widehat{\mathbf{\Lambda}}_l\}}{\partial \mathbf{B}_{l+1}} = \frac{2\mathbf{\Lambda}_l \left(\mathbf{B}_{l+1}\frac{\mathbf{\Lambda}_l}{\mathbf{\Lambda}_{l+1}} - 1\right)}{\left(1+\mathbf{B}_{l+1}\right)^3} \stackrel{!}{=} 0 \quad \Rightarrow \mathbf{B}_{l+1} = \frac{\mathbf{\Lambda}_{l+1}}{\mathbf{\Lambda}_l} \,. \tag{4.87}$$

Checking the second partial derivative

$$\left.\frac{\partial^2 \mathrm{Var}\{\widehat{\mathbf{\Lambda}}_l\}}{\partial \mathbf{B}_{l+1}^2}\right|_{\mathbf{B}_{l+1}=\frac{\mathbf{\Lambda}_{l+1}}{\mathbf{\Lambda}_l}} = \frac{2\mathbf{\Lambda}_l \left(1+\frac{\mathbf{\Lambda}_l}{\mathbf{\Lambda}_{l+1}}\right)}{\left(1+\frac{\mathbf{\Lambda}_{l+1}}{\mathbf{\Lambda}_l}\right)^4} > 0 \tag{4.88}$$

yields that choosing $\mathbf{B}_{l+1} = \frac{\mathbf{\Lambda}_{l+1}}{\mathbf{\Lambda}_l}$ indeed provides the minimum of the mean squared error.

Equations (4.83), (4.85) and (4.87) provide the final estimator

$$\widehat{\mathbf{\Lambda}}_l = \frac{\mathrm{Y}_l + \mathrm{Y}_{l+1}}{1+\frac{\mathbf{\Lambda}_{l+1}}{\mathbf{\Lambda}_l}} \,. \tag{4.89}$$

It has the MSE

$$\mathrm{MSE}\{\widehat{\mathbf{\Lambda}}_l\} = \mathrm{Var}\{\widehat{\mathbf{\Lambda}}_l\} = \frac{\mathbf{\Lambda}_l + \mathbf{\Lambda}_{l+1}}{\left(1 + \frac{\mathbf{\Lambda}_{l+1}}{\mathbf{\Lambda}_l}\right)^2} = \frac{1}{1 + \frac{\mathbf{\Lambda}_{l+1}}{\mathbf{\Lambda}_l}} \mathbf{\Lambda}_l \,. \tag{4.90}$$

Due to the fact that $\frac{1}{1+\frac{\mathbf{\Lambda}_{l+1}}{\mathbf{\Lambda}_l}} < 1$, the variance of the estimator is smaller than the variance $\mathbf{\Lambda}_l$ of the image random variables $\mathtt{Y}_l$. This result is the core of MIDWAI for Poisson noise. It means that it is likely that the estimation is closer to the true mean than the observed realization of the image random variables, therefore reducing the noise.

The ratio of the true values $\frac{\mathbf{\Lambda}_{l+1}}{\mathbf{\Lambda}_l}$ is not known and actually is to be determined by the estimator. As this is a circular argument, the ratio has to be obtained in a different way. This can be done by taking the spatial information into account. A very easy and efficient way is the convolution of both images $\mathbf{Y}_l$ and $\mathbf{Y}_{l+1}$ with a two-dimensional discrete Gaussian kernel. In all experiments, the width N of the quadratic kernel is defined as

$$N = 1 + 8\sigma_{\mathrm{GK}} \,, \tag{4.91}$$

where σ_{GK} is the standard deviation of the Gaussian kernel. σ_{GK} is one scalar parameter of the MIDWAI method. After convolving $\mathbf{Y}_l$ and $\mathbf{Y}_{l+1}$ with the Gaussian kernel, the ratio $\frac{\mathbf{\Lambda}_{l+1}}{\mathbf{\Lambda}_l}$ is defined to be the pixelwise division of the two filtered images.

From the estimator $\widehat{\mathbf{\Lambda}}_l$ (4.89), the denoised image

$$\widehat{\mathbf{Y}}_l = \frac{\mathbf{Y}_l + \mathbf{Y}_{l+1}}{1 + \frac{\mathbf{\Lambda}_{l+1}}{\mathbf{\Lambda}_l}} \tag{4.92}$$

is deduced. It is obtained by an easy and straightforward procedure: The two neighboring two-dimensional images are filtered with the Gaussian kernel and divided pixelwise. Adding one gives the denominator in (4.89). The images $\mathbf{Y}_l$ and $\mathbf{Y}_{l+1}$ are added pixelwise and divided again pixelwise by this denominator.

This procedure can be easily extended to adding more than one image. In case that Z images are to be added, the denoised image is calculated as follows:

$$\widehat{\mathbf{Y}}_l = \frac{\mathbf{Y}_l + \mathbf{Y}_{l+1} + \cdots + \mathbf{Y}_{l+Z}}{1 + \frac{\mathbf{\Lambda}_{l+1}}{\mathbf{\Lambda}_l} + \cdots + \frac{\mathbf{\Lambda}_{l+Z}}{\mathbf{\Lambda}_l}} . \tag{4.93}$$

The number Z of images that are added is the second scalar parameter of MIDWAI. Figure 4.20 graphically shows the procedure of MIDWAI denoising.

4.7.2 MIDWAI for Gaussian Noise

When the major noise component is Gaussian, the assumption that the mean of the random variable equals its variance is no longer valid. Starting from (4.84), however,

$$\mathbf{C}_{l+1} = \frac{\mathbf{\Lambda}_l}{\mathbf{\Lambda}_{l+1}} \tag{4.94}$$

still holds. The variance of the estimator in the case of Gaussian noise is calculated by

$$\mathrm{Var}\{\widehat{\mathbf{\Lambda}}_l\} = \frac{\mathrm{Var}\{\mathtt{Y}_l\} + \mathrm{Var}\{\mathtt{Y}_{l+1}\}\mathbf{B}_{l+1}^2 \left(\frac{(\mathbf{\Lambda}_l)}{\mathbf{\Lambda}_{l+1}}\right)^2}{(1 + \mathbf{B}_{l+1})^2} . \tag{4.95}$$

The next calculation steps are performed analogously to the ones described in the previous subsection. The MIDWAI estimator for Gaussian noise finally results to

$$\widehat{\mathbf{\Lambda}}_l = \frac{\mathtt{Y}_l + \frac{\mathrm{Var}\{\mathtt{Y}_l\}}{\mathrm{Var}\{\mathtt{Y}_{l+1}\}} \frac{\mathbf{\Lambda}_{l+1}}{\mathbf{\Lambda}_l} \mathtt{Y}_{l+1}}{1 + \frac{\mathrm{Var}\{\mathtt{Y}_l\}}{\mathrm{Var}\{\mathtt{Y}_{l+1}\}} \left(\frac{\mathbf{\Lambda}_{l+1}}{\mathbf{\Lambda}_l}\right)^2} . \tag{4.96}$$

The MSE is calculated by

$$\mathrm{MSE}\{\widehat{\mathbf{\Lambda}}_l\} = \mathrm{Var}\{\widehat{\mathbf{\Lambda}}_l\} = \frac{1}{1 + \frac{\mathrm{Var}\{\mathtt{Y}_l\}}{\mathrm{Var}\{\mathtt{Y}_{l+1}\}} \left(\frac{\mathbf{\Lambda}_{l+1}}{\mathbf{\Lambda}_l}\right)^2} \mathrm{Var}\{\mathtt{Y}_l\} \tag{4.97}$$

from which we again see that the variance of the estimator is lower than the one of the original image $\mathtt{Y}_l$. Just like for the Poisson version, the

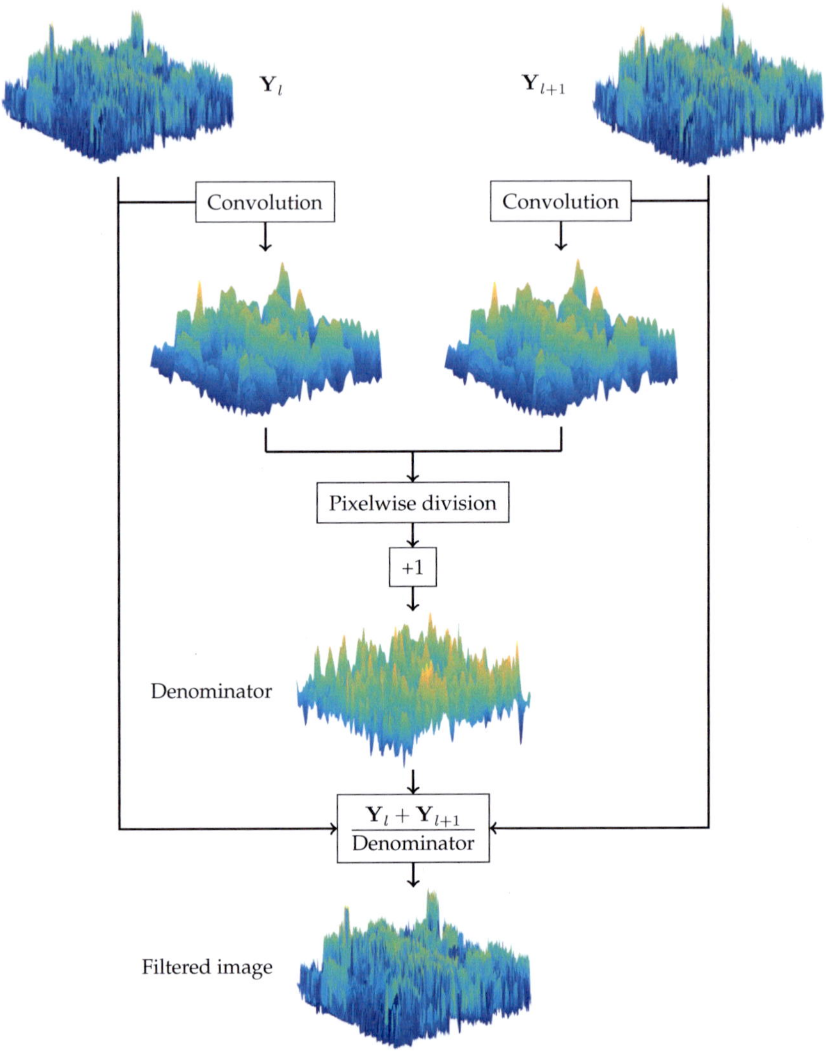

Figure 4.20 MIDWAI Poisson denoising procedure using bands 110 and 111 of the Washington DC Mall image. The SNR increased from 21.56 dB to 24.66 dB. Only two images have been taken into account; the denoising performance increases with an increasing number of images being added weightedly.

mean values are also calculated by convolving the images with Gaussian kernels. The variances $\mathrm{Var}\{\mathtt{Y}_l\}$ and $\mathrm{Var}\{\mathtt{Y}_{l+1}\}$ are estimated from the spatial neighborhood $\mathcal{N}(x, y)$ around the respective pixel at position (x, y) using the sample variance

$$\mathrm{Var}\{\mathtt{Y}_{l,xy}\} = \frac{1}{N^2-1} \sum_{m,n \in \mathcal{N}(x,y)} \left(\mathbf{Y}_{l,mn} - \overline{\mathbf{Y}}_{l,xy}\right)^2 \tag{4.98}$$

$$= \frac{1}{N^2-1} \sum_{m,n \in \mathcal{N}(x,y)} \mathbf{Y}_{l,mn}^2 - \frac{N^2}{N^2-1} \cdot \overline{\mathbf{Y}}_{l,xy}^2 \,. \tag{4.99}$$

In this equation, it is necessary to not use the random variables, but the acquired image values, i.e., their realizations. $\mathbf{Y}_{l,xy}$ denotes the image value at position (x, y) of band l. The neighborhood is defined to the square window of size $N \times N$ around (x, y), which has the same size as the Gaussian filter kernel (4.91). It is very important to note that (4.98) is purely calculated from the pixels inside this square window. It is not feasible to use the mean estimation already obtained by filtering with the Gaussian kernel. The estimations of the mean within the square window and obtained by Gaussian filtering are likely to be different, which, according to (4.99), can lead to negative variance estimations when the mean estimation calculated by Gaussian filtering is too large. Finally, after estimating the variance of all pixels, it is also convolved with the Gaussian kernel (4.91).

According to (4.96), in analogy with (4.92), the denoised image is finally calculated by

$$\widehat{\mathbf{Y}}_l = \frac{\mathbf{Y}_l + \frac{\mathrm{Var}\{\mathtt{Y}_l\}}{\mathrm{Var}\{\mathtt{Y}_{l+1}\}} \frac{\mathbf{\Lambda}_{l+1}}{\mathbf{\Lambda}_l} \mathbf{Y}_{l+1}}{1 + \frac{\mathrm{Var}\{\mathtt{Y}_l\}}{\mathrm{Var}\{\mathtt{Y}_{l+1}\}} \left(\frac{\mathbf{\Lambda}_{l+1}}{\mathbf{\Lambda}_l}\right)^2} \,. \tag{4.100}$$

Analogously to (4.93), it is possible to take Z neighboring images into account instead of just the single band $l + 1$.

The MIDWAI denoising results will be presented and compared with state-of-the-art methods in Sec. 5.3.

5 Results and Comparison

5.1 Images Used for Evaluation

5.1.1 Images Used for Unmixing Experiments

For analyzing the presented methods, several images will be used; some remotely sensed ones and some acquired in a laboratory environment. Table 5.1 provides the parameters of all used images.

Table 5.1 Data of the used hyperspectral images.

Name	Size/pixels	bands L	endmembers R	Type
Urban Detail4	100×100	162	4	remote
Urban Detail6	100×100	162	6	remote
Jasper	100×100	198	4	remote
White Powders	380×320	204	5	lab
Spatial Mixing 10	295×200	206	7	lab
Spatial Mixing 20	147×200	206	7	lab
Spatial Mixing 30	98×200	206	7	lab

The Urban image [64, 160] is a remotely sensed hyperspectral image which captures an urban scene mainly consisting of houses, roads, grass and trees. There are three versions decomposing the image into 4, 5 and 6 pure substances and their abundances, respectively. The first four endmembers are asphalt, grass, tree and roof. The fifth one is soil and the sixth metal. For not having to process the full image every time, two details of size 100×100 are cut out. The first one, Urban Detail4, starts at pixel vertical index 50, horizontal index 100 and only contains the first four endmembers. The second one is located beginning from vertical index 51 and the first horizontal index and consists of all six endmembers.

The Jasper image [64, 160] also was acquired remotely. It has a large part of water in the middle; apart from that, it contains a road, trees and soil.

The datasets acquired in a laboratory environment demonstrate possible applications of spectral unmixing for laboratory and industrial purposes. The White Powders dataset shows how material mixtures can be decomposed into the pure spectra and their abundances. It is a hyperspectral image of the five visually white powders lactose, corn starch, sugar powder, magnesium carbonate and calcium sulfate. These are mixed in different ratios and filled into compartments of an organizer box. This dataset has also been used in [176]. The powders can hardly be distinguished by the human eye, let alone can the mixture ratios be determined. They, however, exhibit characteristic spectral differences in another wavelength range: the original image consisted of 256 bands in the range from 1051 nm–2457 nm in the near-infrared (NIR) and short-wavelength infrared (SWIR) wavelength regions. Out of these 256 bands, the first 20 and last 30 are discarded due to low signal-to-noise ratio. The image was acquired with a Specim SP-SWIR-LVDS-100-N25E SWIR linescan camera. In addition to the image of the scene, black balance and white balance images also have been acquired. For calibration, white balance is performed. The black balance image is recorded with shutter closed and gives an estimate of the dark current in each pixel. The white balance image uses a reflectance standard with known constant reflectance of almost 100 %. After performing the pixelwise calculation [69]

$$\mathbf{y}_i^* = \frac{\mathbf{y}_i - \mathbf{y}_{i,\text{black}}}{\mathbf{y}_{i,\text{white}} - \mathbf{y}_{\text{i,black}}}, \tag{5.1}$$

the image spectra represent relative reflectance at each wavelength, and spatial and spectral illumination variations have been eliminated. The White Powders dataset is calculated with 6 endmembers (five powders plus box plastic), but for evaluation and error calculation, only the five powder spectra are considered. Only the center parts of each compartment are considered for calculating the abundance error (Fig. 5.1). The box plastic and the edge of the compartments where mixing between powder and plastic spectra occurs is left out.

The goal of the images Spatial Mixing 10, Spatial Mixing 20 and Spatial Mixing 30 is to demonstrate another physical effect in hyperspectral image

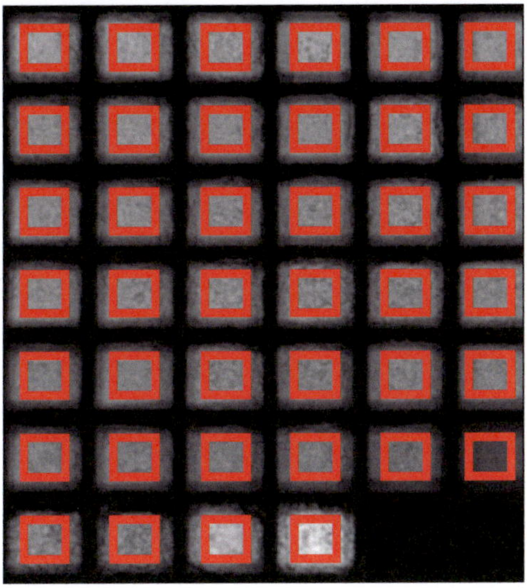

Figure 5.1 White Powders image: regions considered for abundance error calculation.

acquisition and analysis that can be reversed at least partially by application of spectral unmixing. Many industrial sorting machines consist of a moving conveyor belt on which the objects that are to be sorted are placed, see Fig. 1.4. The moving conveyor belt is recorded by line scan cameras which, in combination with the belt motion, produce a full image of the belt surface. Current hyperspectral cameras have rather low line scan rates of 50 to a few hundred lines per second. Although there are other types of cameras with higher rate, they have to sacrifice either spatial or spectral resolution for achieving such high rates. Assuming a scan rate of 100 lines per second and taking typical belt speeds of 3 to 6 meters per second, one spatial pixel covers a distance of 3 to 6 cm along the direction of the belt movement. This means that smaller objects present within these pixels cannot be detected, and subpixel methods such as spectral unmixing have to be used. Applying spectral unmixing provides an answer which percentage of the surface is covered with which kind of object. Although it is not possible to determine the exact position of a foreign object on the belt, at least it can be determined if it is present within the pixel area.

Table 5.2 Numbers of objects placed in the Spatial Mixing base image.

Material	# cropped obj.	# placed obj.	# placed rotated obj.
Grapes	20	30	10
Nuts	16	30	10
Grass pellets	12	7	21
Pepper	9	75	25
Stones	9	1000	1000
Tobacco	6	50	50

This scenario is exactly the same as hyperspectral remote sensing with low spatial resolution.

The three Spatial Mixing images are all based on the same hyperspectral image. This basis image was created by cropping objects from several real hyperspectral images [1] and placing them randomly in a new image. The objects were grapes, nuts, grass pellets, dried pepper rags, pieces of tobacco leaves and small stones. Table 5.2 gives the number of objects for each of the six classes. Note that the background represents the seventh endmember. The objects were selected such that they are relatively small. The pellets are long and narrow, so these and the other objects occur not only in always the same orientation, but also randomly rotated. The whole image can be regarded as a sorting problem where the small stones have to be detected and removed in a subsequent processing step. If one pixel contains a small abundance of stone, its complete area could be sorted out for making sure that no such stones are present in the target products.

The resulting image is of size 2952×200. Figure 5.2 shows how the Spatial Mixing datasets are created from this basis image. Regarding the spatial ground truth, each pixel of the basis image is considered to exclusively belong to one class, i.e., having 100 % abundance of this class. When performing the spatial mixing (Fig. 5.2), mixed pixels result. The object spectra exhibit considerable variability (Sec. 2.1.2). For this reason, the mean spectrum of all spectra belonging to one object class is considered as the ground truth spectrum of this class.

[1] Courtesy of the department Visual Inspection Systems of the Fraunhofer Institute of Optronics, System Technologies and Image Exploitation (IOSB)

Figure 5.2 Creation of the Spatial Mixing datasets. The belt motion is in the vertical direction. All images show the 100th band of the considered hyperspectral image. The basis image is the leftmost image (only the upper fourth is shown), while the next three images have pixels calculated from the average of 10, 20 and 30 pixels along the vertical direction. The number of averaged pixels is repeated in the respective dataset name.

All images described in this section are normalized to have the global mean one, i.e., the mean of all elements of the image matrix $\mathbf{Y}$ is one. The reason is that different scaling of the spectra leads to a more time-consuming search for the optimal parameter μ_{spec}, because the balance between data fidelity and regularizer shifts and a larger range of parameter values has to be investigated. Normalization scales down the range for all images, as it can be expected that the regularizer then has similar values for all images. For this reason, whenever spectra are plotted, no ordinate label is assigned; the spectra magnitude is not encoding relative reflectance anymore.

5.1.2 Images Used for Denoising Experiments

For the MIDWAI Poisson version, hyperspectral fluorescence images of two scenes containing mineral samples have been acquired with the setup described in Sec. 2.3. This setup acquires images with an electron multiplying charge-coupled device (EMCCD) camera (*Andor iXon*$_3$ *897*). As described in Sec. 2.3, the statistics of the EMCCD signal can be approximated very well with the Poisson distribution. Due to the fact that the emitted fluorescence spectra depend on the used ultra-violet (UV) illumination wavelength, the recorded hyperspectral fluorescence images also depend on the illumination. Therefore, there is a fourth image dimension in case a complete three-dimensional image is acquired at various different illumination wavelengths. For this reason, we can use not only the two-dimensional images of the hyperspectral image acquired at a certain wavelength, but use additional images coming from hyperspectral images at different excitation (i.e., illumination) wavelengths.

The excitation wavelength was varied in steps of 20 nm. For evaluating the denoising results, images of six bands with the parameters given in Table 5.3 have been acquired 500 times each. The respective mean of the 500 images is regarded as ground truth due to the unbiasedness of the noise. The excitation wavelength is denoted by λ_{exc} and the emission wavelength by λ_{em}. Table 5.3 additionally gives the configuration of the neighboring images that have also been acquired and can be used for denoising. Only one acquisition of each of the neighboring images was performed instead of 500 as for the ground truth images. The dataset numbers 1 and 2 are images of the first scene and numbers 3–6 are images of the second scene. The number Z of images that are added weightedly is split into two; the first half of the images is taken from the neighboring shorter emission wavelengths, the second half from longer ones, see also Fig. 5.5.

For the Gaussian MIDWAI version, a remotely-sensed hyperspectral image will be used: The HYDICE (HYperspectral Digital Imagery Collection Experiment) image of the Washington DC Mall provided in [101]. The whole image consists of 1280×307 pixels and 191 wavelengths. An image detail of size 251×201 pixels that approximately incorporates the area studied in [155] is considered. It comprises the area from the vertical pixels 550–800 and the horizontal pixels 70–270. All 191 wavelengths are taken into account. The image is degraded artificially by Gaussian noise.

Table 5.3 Parameters of the ground truth images.

Dataset number	1	2	3	4	5	6
$\lambda_{\text{exc}}/\text{nm}$	360	380	380	380	400	400
$\lambda_{\text{em}}/\text{nm}$	550	558	558	602	558	602
Adjacent images: $\lambda_{\text{exc}}/\text{nm}$	340, 380	360, 400	340, 360, 400, 420	340, 360, 400, 420	380, 420	380, 420
Adjacent images: $\lambda_{\text{em}}/\text{nm}$	450 – 650	458 – 658	458 – 658	502 – 702	458 – 658	502 – 702

5.2 Comparison Criteria

For evaluating the unmixing performance of different methods, error measures will be used. In these measures, the matrices $\widehat{\mathbf{M}}$ and $\widehat{\mathbf{A}}$ are the matrices calculated by the unmixing method, while $\mathbf{M}$ and $\mathbf{A}$ denote the ground truth.

The abundance root-mean-square error (ABU RMSE) is calculated by

$$\text{ABU RMSE}(\mathbf{A}, \widehat{\mathbf{A}}) = \frac{1}{R}\sum_{r=1}^{R}\left(\frac{1}{I}\sum_{i=1}^{I}(a_{ri} - \widehat{a}_{ri})^2\right)^{\frac{1}{2}} \tag{5.2}$$

and the endmember RMSE (EM RMSE) by

$$\text{EM RMSE}(\mathbf{M}, \widehat{\mathbf{M}}) = \frac{1}{R}\sum_{r=1}^{R}\left(\frac{1}{L}\sum_{l=1}^{L}(m_{lr} - \widehat{m}_{lr})^2\right)^{\frac{1}{2}} . \tag{5.3}$$

Another error measure for the endmember matrix is the spectral angle (SA, see also Sec. 2.2.3) defined by

$$\mathrm{SA}(\mathbf{M},\hat{\mathbf{M}}) = \sum_{r=1}^{R} \arccos\left(\frac{\sum_{l=1}^{L} m_{lr}\hat{m}_{lr}}{\sqrt{\sum_{l=1}^{L} m_{lr}^2}\sqrt{\sum_{l=1}^{L} \hat{m}_{lr}^2}}\right). \tag{5.4}$$

The spectral angle neglects constant factors the spectra might be multiplied with, while the RMSE takes them into account. According to [36], endmember SA and RMSE are highly correlated. For this reason, depending on the analysis that is to be assessed, the most appropriate endmember measure will be used.

In case the abundance and/or endmember matrix ground truth is not known, the IMAGE RMSE can be used for assessing how well the measured image is approximated by the estimated abundances and endmembers. The IMAGE RMSE is calculated by

$$\text{IMAGE RMSE}(\mathbf{Y},\widehat{\mathbf{Y}}) = \left(\frac{1}{IL}\sum_{i=1}^{I}\sum_{l=1}^{L}\left(y_{li}-\widehat{y}_{li}\right)^2\right)^{\frac{1}{2}}. \tag{5.5}$$

For the denoising experiments, the bandwise peak signal-to-noise ratio (PSNR) will be used. For band l of the denoised hyperspectral image, it is calculated by

$$\mathrm{PSNR}_l = 10\log_{10}\left(\frac{\max\left(\mathbf{Y}_{\mathrm{gt},l,xy}\right)^2}{\frac{1}{XY}\sum_{x=1}^{X}\sum_{y=1}^{Y}\left(\mathbf{Y}_{\mathrm{gt},l,xy}-\widehat{\mathbf{Y}}_{l,xy}\right)^2}\right), \tag{5.6}$$

where $\mathbf{Y}_{\mathrm{gt},l}$ denotes the ground truth band l. The mean PSNR (MPSNR) is the mean PSNR of all bands. For denoising, but also for the 3DDCT-NIMF discussion, the SNR is calculated by

$$\mathrm{SNR}_l = 10\log_{10}\left(\frac{\sum_{x=1}^{X}\sum_{y=1}^{Y}\left(\mathbf{Y}_{\mathrm{gt},l,xy}\right)^2}{\sum_{x=1}^{X}\sum_{y=1}^{Y}\left(\mathbf{Y}_{\mathrm{gt},l,xy}-\widehat{\mathbf{Y}}_{l,xy}\right)^2}\right). \tag{5.7}$$

For the 3DDCT-NIMF consideration, this formula is adapted to the respective area of the 3D data structure.

5.3 MIDWAI Denoising

The evaluation of both the Gaussian and the Poisson version of the MIDWAI (multidimensional image denoising by weighted addition of images) denoising procedure will be presented in this section. The MIDWAI results will be compared with the results of the spectral–spatial adaptive hyperspectral total variation (SSAHTV) denoising algorithm given in [155] and the block-matching and 4D filtering (BM4D) algorithm [91]. BM4D is an extension of the powerful block-matching and 3D filtering (BM3D) algorithm [34] to three-dimensional images. The source code of SSAHTV has kindly been provided by the authors, while the BM4D implementation is provided online [66].

Table 5.4 gives the denoising results with the MIDWAI procedure applied to the fluorescence emission images in three different configurations:

- The mean estimation is calculated according to Section 4.7.1 (convolution with Gaussian kernel).
- Alternatively, the mean estimation is obtained by applying other filters such as BM3D to each of the original 2D images (so-called prefiltering).
- After calculating the MIDWAI denoised images, it is possible to remove the remaining noise by applying generic 2D filters band by band, in this case the BM3D filter.

The BM4D and SSAHTV results are also given. For these algorithms, the Anscombe transform [92] was applied first to convert the Poisson noise to Gaussian noise, as both algorithms are not explicitly designed for Poisson noise. From Table 5.4, it can be seen that MIDWAI denoising with the standard Gaussian convolution provides slightly better results than BM3D prefiltering. In spite of its simplicity, the Gaussian convolution provides PSNR results in the same range as BM4D, while SSAHTV performs slightly worse. MIDWAI with Gaussian convolution and weak BM3D postprocessing provides the best denoising results. Figures 5.3 and 5.4 show the denoised images. It can be seen that the denoising with MIDWAI removes the noise almost completely (only little noise is left), while there are no artifacts such as overly smoothed areas. Note that the images filtered by

Gaussian convolution are only used for weighting the images, but then, the original noised images are added. This means that no blurring of edges results from applying MIDWAI. Another detail image from dataset 1 is shown in Fig. 5.6. It has a stronger signal component than the excerpt from dataset 4.

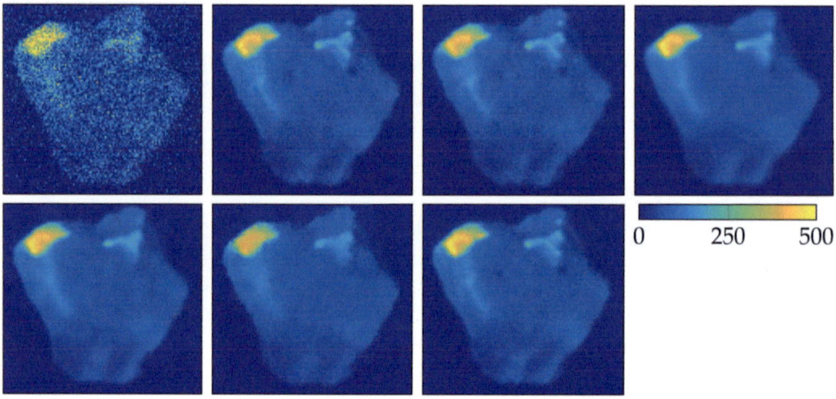

Figure 5.3 Dataset 4: Detail of one of the mineral samples in the image (108×123 pixels). On average, only 6–7 photons per pixel are registered. First row: Noisy image, MIDWAI with Gauss prefiltering, MIDWAI with BM3D prefiltering, MIDWAI with Gauss prefiltering and BM3D postprocessing. Second row: BM4D, SSAHTV, ground truth. Dark blue corresponds to the lowest pixel values, while yellow corresponds to the largest. Note that the MIDWAI images without postprocessing preserve edges really well and are less blurry. Details such as the two dark spots and the light spot at the very top of the image can still be recognized. These details are almost completely lost in the BM4D and SSAHTV images.

As mentioned before, hyperspectral fluorescence images are acquired, which means that for each excitation wavelength, there is a complete hyperspectral image. Due to the fact that MIDWAI adds 2D images, regardless of the dimension they are located in, all the 4D data can be incorporated. Figure 5.5 shows how the 2D subimages from hyperspectral images at neighboring images can be taken into account. This time, the central excitation wavelength and the two neighboring ones are used. Z defines the number of all images from one excitation that is used. As three excitations are used, the total number of images taken into account

Table 5.4 Top table: Optimal PSNR values of the denoised hyperspectral fluorescence images (DS: dataset) in dB. Only the emission images of one excitation wavelength (see Fig. 5.5) have been taken into account. The prefiltering method (*pre* for short) indicates how the mean estimation was obtained: convolution with Gaussian kernel (MIDWAI as explained in Sec. 4.7.1) or BM3D filtering of the two-dimensional images. As the MIDWAI filtered image is not completely noise-free, it can be filtered with 2D filters with low filtering strength afterwards (BM3D column: MIDWAI with Gaussian prefiltering and BM3D after MIDWAI denoising). This leads to the best results among all methods. The bottom table shows the parameter values which provided the optimal results given in the top table. For BM4D, $\gamma = 0$ indicates that the internal noise estimation of the algorithm is active.

DS	Noisy image		MIDWAI Gauss pre	MIDWAI BM3D pre	MIDWAI BM3D	BM4D	SSAHTV
	SNR	PSNR	PSNR	PSNR	PSNR	PSNR	PSNR
1	14.71	31.83	45.19	44.96	**47.81**	46.93	43.00
2	15.60	32.58	45.43	45.23	**47.53**	45.08	43.50
3	12.53	33.48	47.92	46.77	**50.62**	47.15	44.10
4	9.25	30.72	45.80	44.94	**48.95**	47.52	43.01
5	13.27	34.48	48.79	47.71	**51.44**	47.58	44.69
6	10.08	31.76	46.94	46.04	**50.22**	48.12	43.64

DS	Gauss pre		BM3D pre	BM3D		BM4D	SSAHTV
	$Z/2$	σ_{GK}	$Z/2$	$Z/2$	σ_{GK}	γ	γ
1	21	6	27	16	10	0.005	40
2	24	4	25	17	6	0	40
3	30	7	30	21	8	0.005	40
4	23	11	30	11	20	0.005	30
5	30	7	30	30	7	0.005	40
6	24	7	30	13	16	0.005	30

amounts to $3Z$. Table 5.5 gives the denoising results. It can be seen that incorporating the neighboring excitation wavelengths leads to a PSNR improvement of 2–3 dB, cf. Tab. 5.4.

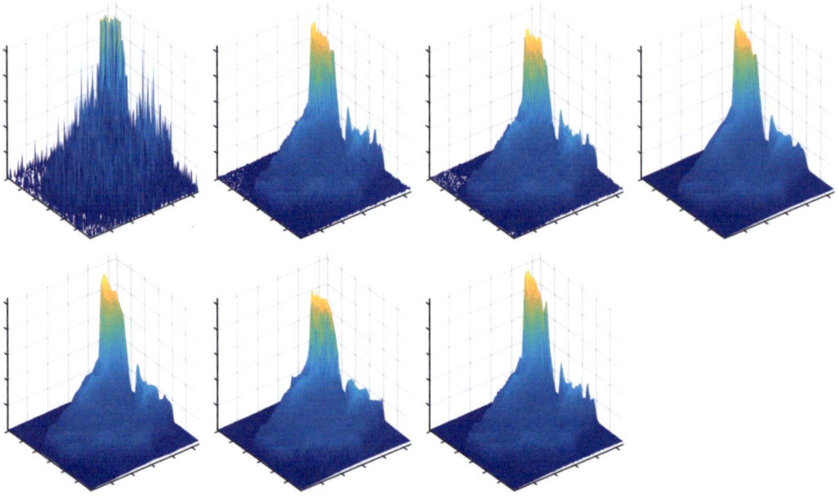

Figure 5.4 Surface plots of the images in Fig. 5.3. The vertical axis shows the image values and is scaled from 0 to 500. First row: Noisy image, MIDWAI with Gauss prefiltering, MIDWAI with BM3D prefiltering, MIDWAI with Gauss prefiltering and BM3D postprocessing. Second row: BM4D, SSAHTV, ground truth.

It should be emphasized that the good quality of the denoised images is not due to random selections of the parameters Z and σ_{GK} that lead to such good results. In fact, the outcome of MIDWAI is pretty robust against changes in the parameters and the PSNR values change continuously with varying Z and σ_{GK}. Figure 5.7 gives the PSNR values for all datasets denoised by MIDWAI with Gaussian convolution (no postprocessing) when only one excitation wavelength is considered.

5.3.1 Gaussian Noise

For evaluating MIDWAI in the presence of Gaussian noise, the Washington DC Mall image is degraded by additive white Gaussian noise (AWGN) with an SNR of 10, 20 and 30 dB. The applied SNR value is the same for the 2D images at all bands. Furthermore, in a fourth variation, the bandwise SNR is selected randomly from the range 15–25 dB to evaluate the behavior

Table 5.5 PSNR values (in dB) of the denoised hyperspectral fluorescence images when the images of neighboring excitation wavelengths (see Fig. 5.5) are also taken into account. The used prefiltering method is the convolution with Gaussian kernel. Note the improvement of 2–3 dB in comparison with Table 5.4.

Dataset	MIDWAI Gauss			MIDWAI BM3D		
	PSNR	Z/2	σ_{GK}	PSNR	Z/2	σ_{GK}
1	47.40	15	11	49.11	10	20
2	47.44	16	7	49.06	12	10
3	50.33	30	8	51.77	14	15
4	48.06	19	17	49.77	9	20
5	51.07	30	8	52.28	16	12
6	49.24	19	11	51.02	9	20

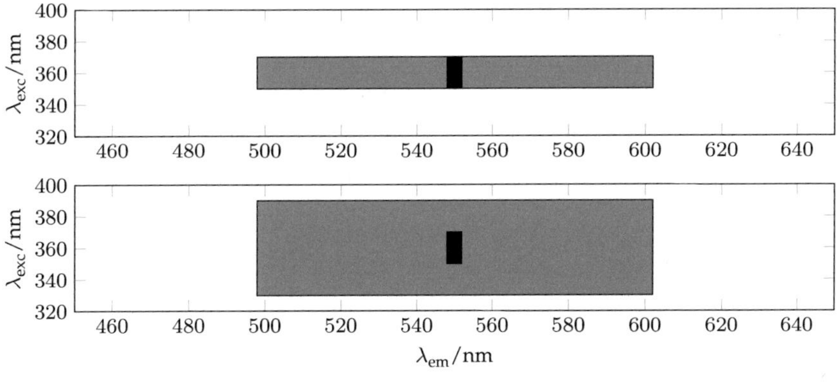

Figure 5.5 Dataset 1: Addition scheme of the neighboring images that are taken into account when only images of the same excitation wavelength (top, denoising results in Tab. 5.4) and of additional neighboring excitation wavelengths are used (bottom, results in Tab. 5.5). The image that is to be denoised is shown in black, the images that are additionally taken into account are marked gray.

of MIDWAI when the bandwise SNR is not constant. Table 5.6 shows the MPSNR denoising results as well as the calculation time of all algorithms. For the Gaussian MIDWAI method, the parameters Z and σ_{GK} which lead

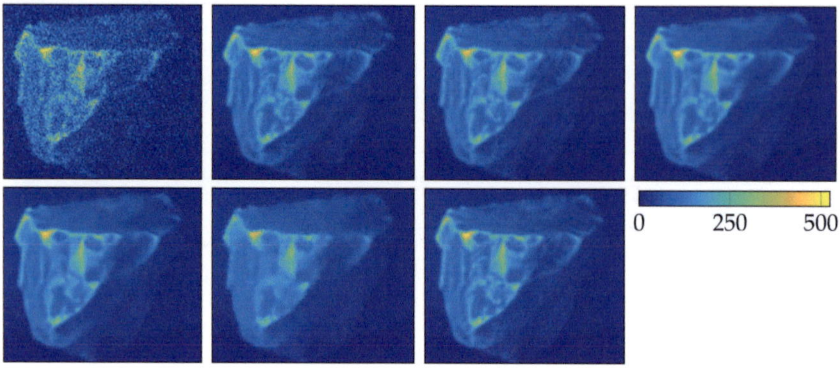

Figure 5.6 Dataset 1: Detail of one of the mineral samples in the image (124×152 pixels). On average, only about 15 photons per pixel are registered. First row: Noisy image, MIDWAI with Gauss prefiltering, MIDWAI with BM3D prefiltering, MIDWAI with Gauss prefiltering and BM3D postprocessing. Second row: BM4D, SSAHTV, ground truth. Dark blue corresponds to the lowest pixel values, while yellow corresponds to the largest. Again, the MIDWAI images without postprocessing preserve details really well and are less blurry. Only a small amount of noise is left. Note also the denoising quality of the darker right side of the sample.

to the highest MPSNR are chosen. These are not the optimal ones for each individual band, but interestingly, the mean of the optimal PSNR values for all bands at all tested parameter configurations at 10 dB SNR amounts to 34.54. This is only about 1 dB more (cf. Tab. 5.6) and shows that it is feasible to select the same parameters for all bands. MIDWAI Gauss provides not only very good denoising results, but performs about 25 times faster than BM4D.

Figure 5.8 shows the bandwise PSNR for all considered algorithms. In general, the MIDWAI and BM4D curves have a similar shape. For assessing the visual quality of the denoised images, Figs. 5.9 and 5.10 exemplarily show the denoised band 20 of the Washington DC Mall image. In general, when both the BM4D and MIDWAI images result in a rather poor approximation of the original image, the BM4D results are overly smoothed, while the MIDWAI images still contain noise of low amplitude or have a small offset compared to the ground truth. For visualizing this effect, Fig. 5.11 shows the denoising results along one image line.

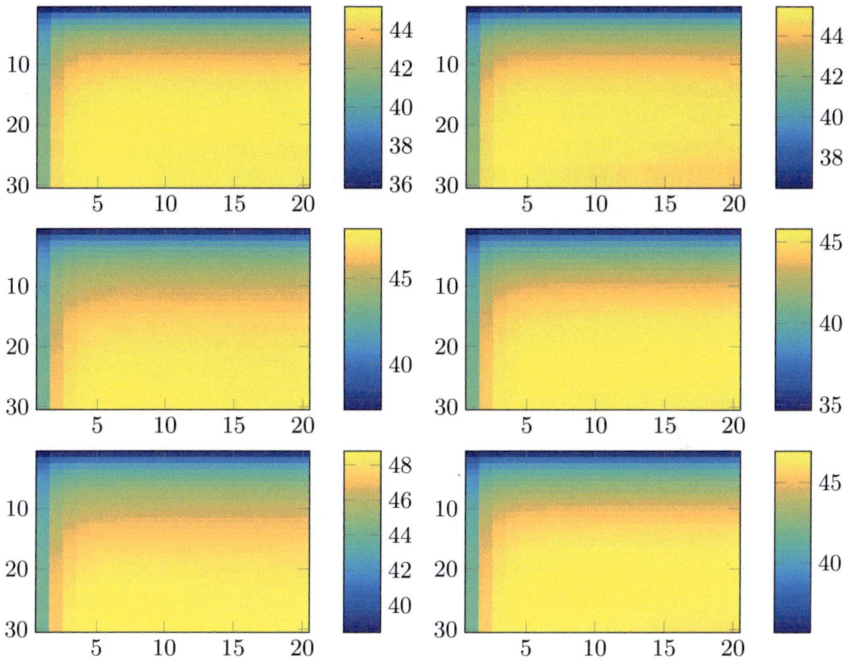

Figure 5.7 PSNR values in dB of all datasets (1–6 from top left to bottom right) in dependence of σ_{GK} (x-axis) and Z (y-axis). When $\sigma_{\mathrm{GK}} = 6$ and $Z = 25$, the mean PSNR of all datasets amounts to 46.56 dB.

5.3.2 Discussion

Although MIDWAI is a denoising method based on a straightforward principle, it shows very good denoising results that can compete with sophisticated state-of-the-art denoising methods. The main characteristic of its denoising results is extraordinary preservation of details without introducing artifacts such as the staircase effect known from conventional TV, merging regions with very similar pixel values or oversmoothing. When it provides images of less quality than other algorithms, this is mainly due to the fact that the noise level has not been reduced sufficiently, but in general, no artifacts are introduced into the image. The only effect that appears when the parameters are not chosen suitably is that there is

Table 5.6 MPSNR values in dB and calculation time of all algorithms applied to the Washington DC Mall image. For all noise configurations, MIDWAI with Gauss prefiltering provides similar results as BM4D, in most cases even slightly better ones. The MIDWAI calculation, though, runs about 25 times faster than BM4D. For BM4D, σ was set to 0, i.e., the noise estimation of the algorithm is active. The used software was *MATLAB* running on a 3.5 GHz i7 processor with 16 GB RAM.

Noise	Noisy im.	MIDWAI Gauss pre				BM4D		SSAHTV		
	MPSNR	MPSNR	Z/2	σ_{GK}	t/s	MPSNR	t/s	MPSNR	γ	t/s
10 dB	21.58	**33.22**	14	18	**25.7**	32.31	494.9	27.45	2	93.6
20 dB	31.58	**39.60**	6	20	**18.7**	39.50	498.8	34.48	10	21.6
30 dB	41.58	45.69	2	21	15.0	**46.77**	500.4	42.50	50	**11.2**
var.	32.07	**39.46**	6	21	**19.7**	39.35	498.0	34.62	10	20.8

an offset between noisy and denoised image. Such an effect also occurs when the spectral values from one image to the others differ greatly (i.e., neighboring images do not exhibit sufficient similarity), but this effect was very small in all experiments that were conducted. If it occurred, the details were still well preserved and the discrimination between details is easy; only a small intensity difference occurred, see Fig 5.11.

One question is how to deal with 2D images that are close to one end of the image, i.e., the first or last bands. For the very first and very last band, it is only possible to weightedly add images of longer or shorter wavelengths, respectively, and not of both wavelengths. One possibility to deal with this problem is to take the respective number of images from the other end of the wavelength range, i.e., for the first band, bands from the long-wavelength end of the image are also used. It turns out that this is not a feasible method if the intensity values at both ends of the wavelength range vary greatly. This is the case with the Washington DC Mall image. If such a cyclic continuation is used, the PSNR values in Fig. 5.8 will decrease much more towards the spectral ends of the image. Therefore, at the shortest and longest wavelengths, only a reduced number of images is used, see Fig. 5.12. This holds for all conducted experiments.

For determining the parameter values, it can be roughly said that for images with high noise amplitudes, more neighboring bands have to be

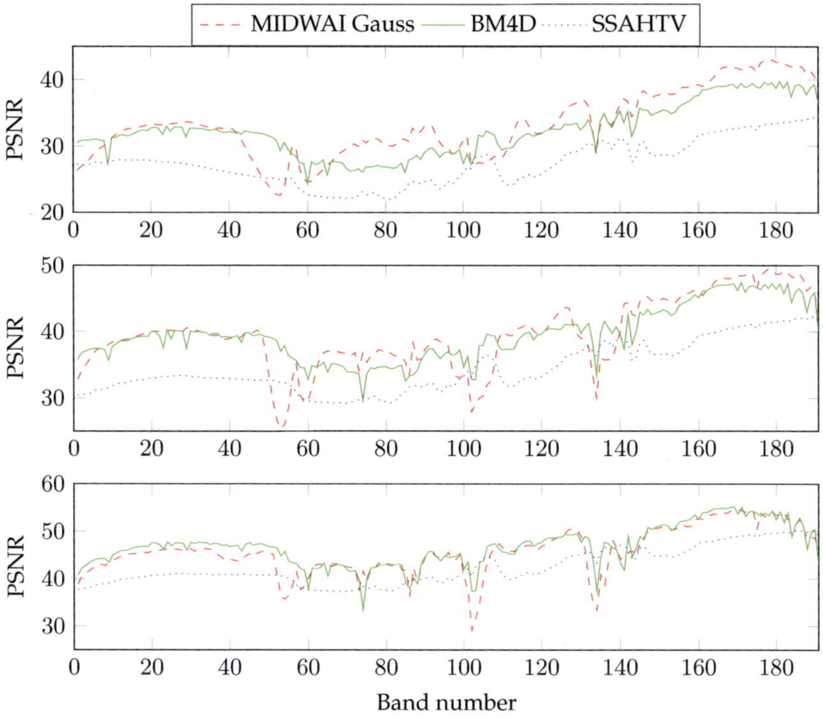

Figure 5.8 Bandwise PSNR of the denoised Washington DC Mall image. Top: noise level 10 dB, middle: noise level 20 dB, bottom: noise level 30 dB.

incorporated than in the case of low noise level (see Table 5.6). One very good indicator is the mean of the noisy and the denoised images. As mentioned above, it might be possible that there is an offset between the noisy and denoised image. For this reason, when the mean of the noisy and the mean of the denoised image differ too much, the parameter values should be adapted.

The computational complexity of the MIDWAI procedure can be deduced from the complexity of the convolution of an image of size $X \times Y$ with a kernel of size $N \times N$. Such a convolution is of complexity $\mathcal{O}(XYN^2)$ (depending on the calculation scheme and hardware, there are faster algorithms for performing the convolution of images with kernels, e.g.,

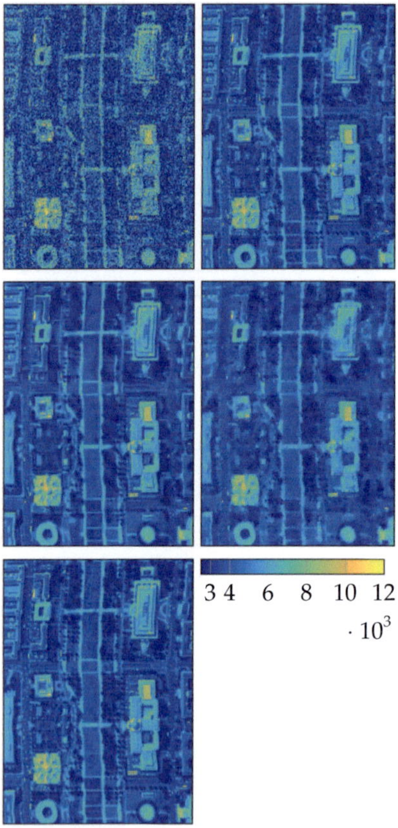

Figure 5.9 Band 20 of the Washington DC Mall image at 10 dB SNR. Top left: noisy image, top right: MIDWAI Gauss, middle left: BM4D, middle right: SSA-HTV, bottom: Ground truth.

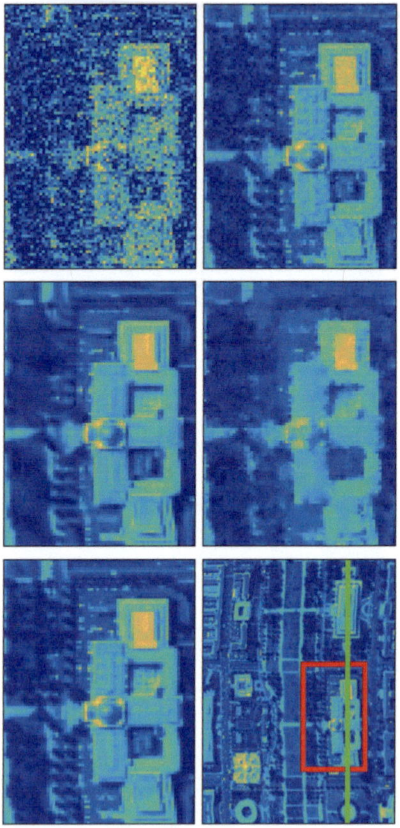

Figure 5.10 Detailed region of band 20 of the Washington DC Mall image at 10 dB SNR. Top left: noisy image, top right: MIDWAI Gauss, middle left: BM4D, middle right: SSAHTV, bottom left: Ground truth, bottom right: image region (red rectangle) and line of interest (green, see Fig. 5.11).

algorithms making use of the fast Fourier transform). For MIDWAI Poisson, the convolution has to be applied to all L bands: $\mathcal{O}(XYN^2L)$. The

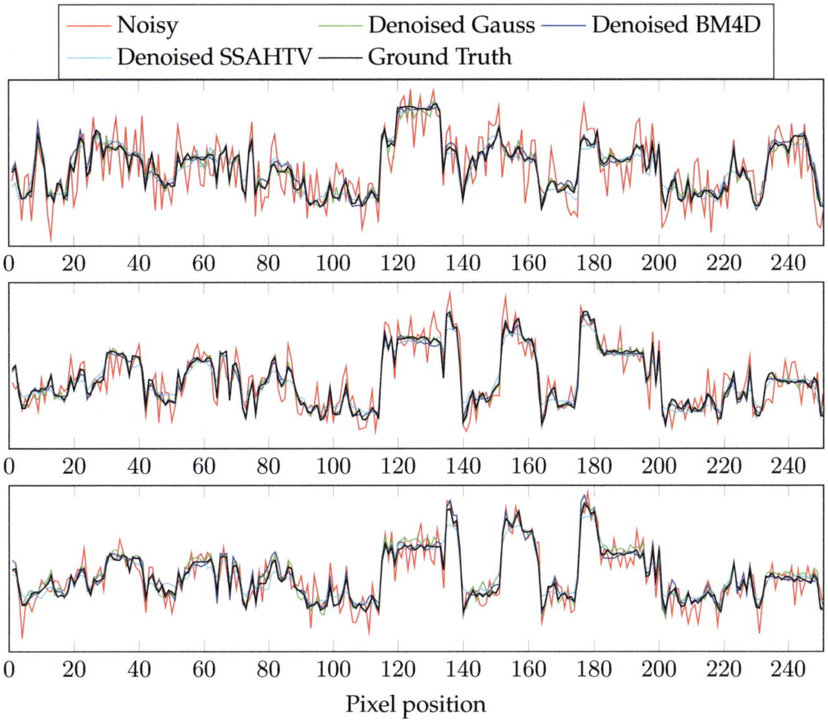

Figure 5.11 Image line that is marked green in Fig. 5.10. Top: band 20, middle: band 178, bottom: band 191. Note how well the edges are preserved. Band 191 has the lowest SNR; the shape of the line is well approximated; however, a small offset is introduced by MIDWAI in some areas, e.g., between pixels 115 and 135.

subsequent divisions and additions according to (4.93) do not increase the complexity.

For MIDWAI Gauss, the only difference is that the pixelwise variances have to be calculated first. For calculating the variance at each pixel's position according to (4.98), the complexity amounts to $\mathcal{O}(XYN^2)$, as for the sum $\sum_{m,n\in\mathcal{N}(x,y)} \mathbf{Y}_{l,mn}^2$, N^2 multiplications and N^2-1 additions have to be calculated. Therefore, the complexity of the variance calculation is the same as the one of calculating the convolution, and the overall complexity of MIDWAI Gauss is the same as the one of MIDWAI Poisson.

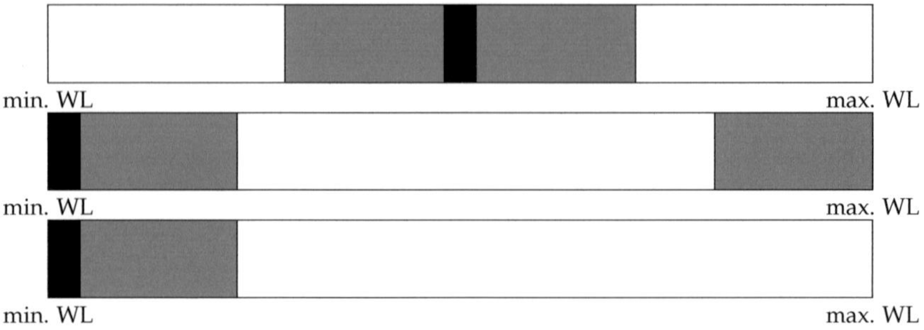

Figure 5.12 Used addition pattern in case of a central band (top) and an edge band (middle). The image that is to be denoised is shown in black and the incorporated neighbor images in gray. When the middle scheme is used, there is probably a loss in denoising quality at the edge bands, as bands with quite different intensities might be mixed. Therefore, these bands are left out (bottom).

In general, the MIDWAI denoising quality is in the range of the top state-of-the-art denoising algorithms. MIDWAI performs slightly better than BM4D, while BM4D has a time demand that is multiple times the one of MIDWAI. The images denoised by MIDWAI can be postprocessed by 2D denoising algorithms with low denoising strength. MIDWAI rapidly provides images with little noise content, therefore it is a suitable workflow to apply methods that specialize in removing small amounts of noise from the whole image such as the ones given in [30] and [106] after MIDWAI. Further improvements of the MIDWAI algorithm could comprise the analysis of the noise that is left in the image for developing adapted postprocessing algorithms. Methods for estimating the bandwise noise level that are, e.g., contained in the maximum noise fraction calculation [52] can be used for estimating the bandwise noise level and adapting the parameters Z and σ_{GK} for each band separately. However, the results we obtained show that this would only lead to a slight improvement.

Apart from the good denoising quality and fast calculation time, MIDWAI can be applied to images of more than three dimensions. In this case, the multidimensional redundancies can be exploited. This has been shown exemplarily with four-dimensional hyperspectral fluorescence images.

5.4 UNCLI

The UNCLI (unmixing of clustered image) unmixing method from Sec. 4.2 has been applied to all datasets described in Sec. 5.1.1. The remotely sensed images are analyzed in three configurations: no noise and AWGN with 20 dB and 30 dB SNR, respectively. UNCLI's main goal is to provide unmixing errors of roughly the same quality as full image unmixing, but in shorter calculation time. The following unmixing scenarios will be analyzed:

- SISAL: this scenario is supervised, i.e., it is based on endmember extraction from the clustered image using the SISAL algorithm (see Sec. 3.2.1). The result of this first step is an endmember estimate; the abundance estimate is obtained by subsequent least-squares calculation (*MATLAB*'s lsqlin() function with the active-set option) from the objective function (3.1) with fixed endmember matrix. It must be remarked that unless otherwise stated, SISAL's parameter τ for the clustered images is chosen to the one that provides the optimal endmember estimate when applied to the unclustered image. The fact that the clustered image has a much smaller image matrix means that there might be better parameter choices for the clustered images. This will also be investigated.

- N-FINDR: similarly to the SISAL scenario, the endmembers are estimated from the respective clustered image using N-FINDR (see Sec. 3.2.1), followed by least-squares abundance estimation. The N-FINDR algorithm does not have any relevant parameters that should be adapted depending on the image size.

- MVC: unsupervised scenario, i.e., both endmembers and abundances are calculated in one procedure. The calculation is performed by alternating between abundances and endmembers using the respective clustered image. MVC spectral regularization (see Sec. 3.2.2) is applied. The regularization parameter μ_{spec} is chosen optimally for the unclustered image; in many cases, this value is also used for the clustered images. A value of μ_{spec} more feasible for the clustered images is also investigated.

To allow for a fair comparison, MVC's initial endmember matrix is chosen randomly, i.e., by randomly selecting R pixel spectra from the unclustered image. This holds for both the clustered and the unclustered images.

- MiniDisCo: same clustering procedure as in the MVC scenario, but using MiniDisCo (see Sec. 3.2.2) as unmixing method instead of MVC. In accordance with the original publication [63], the endmember matrix is set to a matrix containing only zeros.

5.4.1 Supervised Unmixing – UNCLI Clustering Results

The evaluation of the clustering will be analyzed by means of the supervised SISAL and N-FINDR scenarios. This provides the opportunity to not only analyze the segmentation maps, i.e., the spatial maps showing how the image pixels have been merged by the clustering, but also to assess how clustering affects the unmixing quality. This can be done by analyzing the abundance and endmember errors resulting from the unmixing using SISAL or N-FINDR, respectively.

At first, the Urban Detail6 dataset with 30 dB SNR is considered. Each of the clustering methods described in Sec. 4.2.2 is applied to the image and unmixing according to the SISAL scenario is performed. Figure 5.13 shows the segmentation maps that lead to the lowest ABU RMSE among all used parameter combinations of each method. Note that in spite of the noise, the spatial structure is sufficiently preserved and can be observed in the segmentation maps. The number of clusters in the best segmentation maps is 1350 for SP, 2500 for LA, 256 for k-means, 1000 for RHSEG and 66 for ISODATA. The optimal RHSEG result is obtained for $sw = 1$. Note that the number of pixels in the original image, 10000, is greatly reduced with all clustering methods. From the figure, it can be seen that the optimal SP clustering merges relatively large image regions into one cluster, while the optimal RHSEG and k-means segmentation maps put more emphasis on the spectral similarity, i.e., they merge pixels that are not necessarily spatially adjacent. For the considered image, ISODATA's optimal clustering represents a compromise between spatial and spectral predominance.

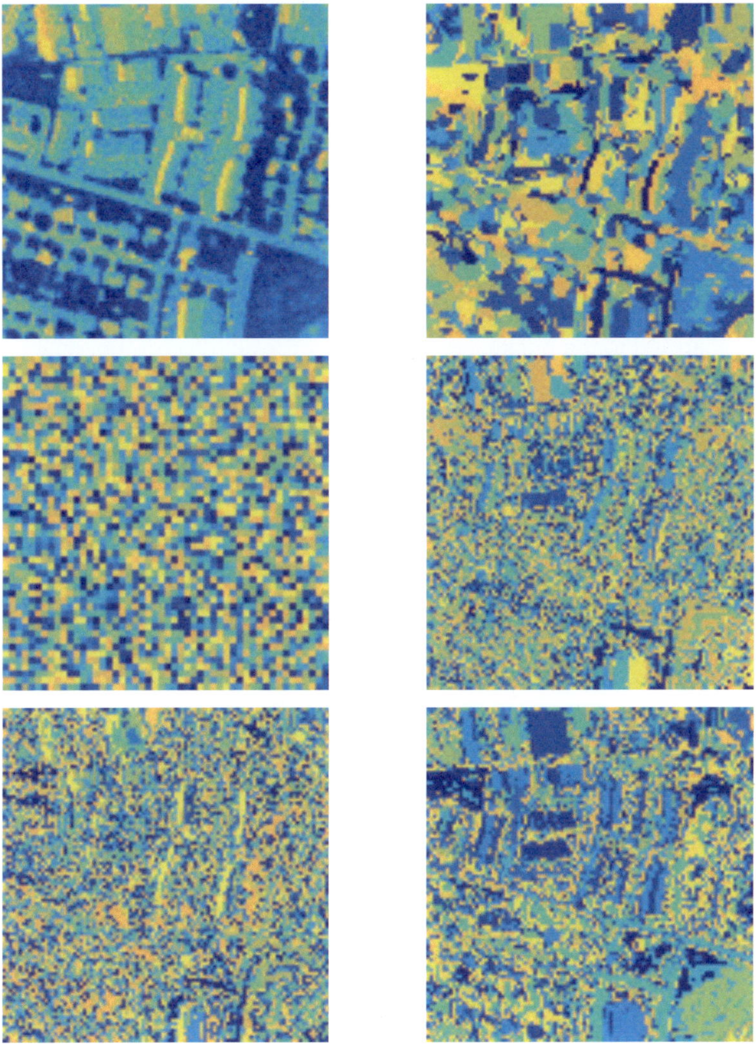

Figure 5.13 SISAL scenario: segmentation maps of the Urban Detail6 dataset that provide the optimal unmixing results within each clustering method. Every cluster represents a cluster. Top row: band 1 of the original image with 30 dB SNR (left), SP segmentation (right). Middle row: LA (left), k-means (right), bottom row: RHSEG (left), ISODATA (right).

5.4.2 Supervised Unmixing – UNCLI Unmixing Results

Figure 5.14 depicts the unmixing results (abundance and endmember errors, see Sec. 5.2) of the noiseless Urban Detail6 image in the N-FINDR scenario for both clustered and unclustered images. Note that the image is decomposed almost perfectly when using the original, unclustered image (the error values are close to zero). For this reason, it is hard for the clustered images to outperform this result; however, the error difference is decreasing with increasing number of clusters, and RHSEG with a high cluster number provides similar results as the unclustered image. When noise is added to the image, see Fig. 5.15 for the analysis of the same image with 30 dB SNR, unmixing the clustered images indeed provides better results than the unclustered image for quite a large number of clusters. Note that especially the endmember estimation greatly benefits from the clustering. The positive effect of spatial information on endmember extraction has already been studied and described in, e.g., [113], where the automated morphological endmember extraction (AMEE) algorithm is presented. Also, a spatial preprocessing algorithm (SPP) [93] has been presented that exploits the spatial information in a first step; in the second step, any spectral EEA can be applied to the preprocessed data.

Comparing the results of UNCLI applied to the Urban Detail6 image without noise (Fig. 5.14) and with 30 dB SNR (Fig. 5.15), it becomes obvious that this method is beneficial when dealing with noisy images. This is most likely due to the averaging effect: each cluster spectrum is calculated by averaging all the pixel spectra belonging to this respective cluster. This means that zero-mean noise such as AWGN is reduced. At a very low number of clusters, the error measures are quite high, and decrease with increasing number of clusters. At a specific number of clusters, the error measures increase again. This effect is especially obvious when RHSEG clustering is used and can be explained as follows: when too few clusters are considered, the spatial resolution of the clustered image is too low, so no acceptable unmixing results can be obtained. At a certain number of clusters, the original image can be represented very well by the clusters, and the noise is mostly eliminated. Further increasing the number of clusters means that only few pixels are merged, so the clustered image does not significantly differ from the original noisy one, meaning that the unmixing results also will be similar. Note that a cluster number of

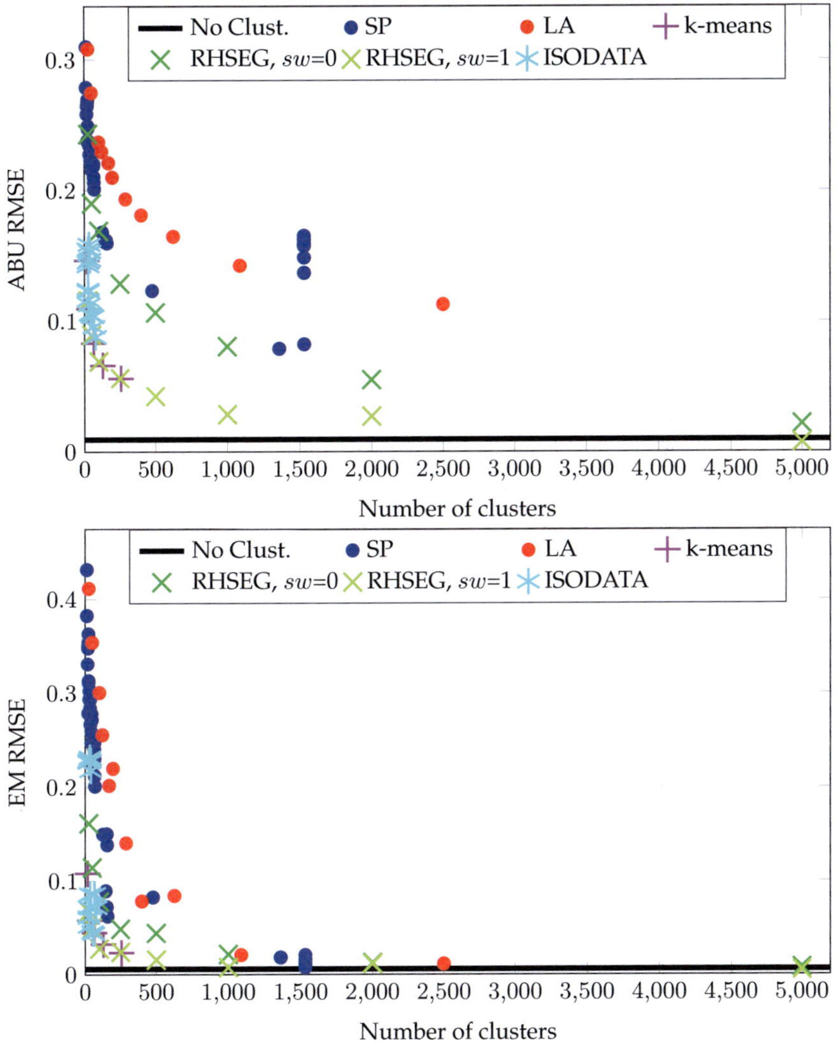

Figure 5.14 Noiseless Urban Detail6 unmixing with N-FINDR endmember initialization. The thick black line represents the results obtained with the unclustered image. All other data points are calculated by UNCLI with different clustering methods. The results of the clustered images become better with increasing cluster number, however, the low error values of the unclustered image are hard to beat.

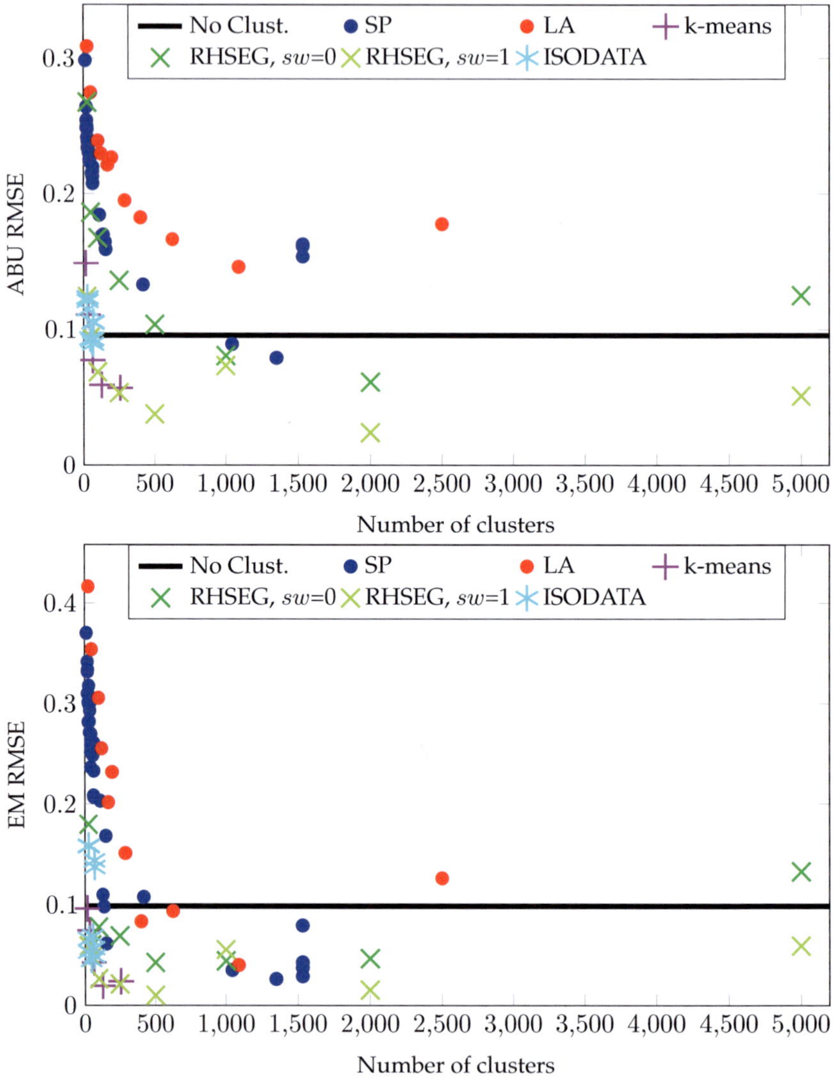

Figure 5.15 Urban Detail6 (30 dB SNR) unmixing with N-FINDR endmember initialization. The thick black line represents the results obtained with the unclustered image. In the presence of noise, UNCLI yields for many clustered images better unmixing results than full image unmixing.

5000 means for the Urban Detail6 image that each cluster on average only consists of two pixels.

For the Urban Detail6 image with larger noise component, 20 dB SNR, the effect already seen at 30 dB noise level becomes even larger; the clustered images provide an even larger improvement in abundance and endmember estimation. The plot for this noise level is omitted.

From the analysis of the Urban Detail6 image, it can be seen that UNCLI leads to an unmixing quality improvement (i.e, lower error values) in the presence of noise; the more noise, the larger the improvement. For each noise level, the best results are obtained for a feasible number of clusters. If it is too low, spatial details are lost; if it is too high, the clustered image is almost the unclustered one, therefore the clustering does not have any impact. Table 5.7 shows the results for the other considered images.

5.4.3 Supervised Unmixing – Algorithm Dependence

Having seen that UNCLI improves the unmixing results in the presence of noise, it is interesting to take a closer look at the respective clustering methods and their parameter selection.

For the RHSEG clustering method, the parameters sw and the chosen distance measure are the most important ones. Again using Urban Detail6 without noise, their effect is shown in Fig. 5.16. The outcome hardly depends on the used distance measure Euclidean distance (ED) or spectral angle (SA), however, the spectral weight parameter sw significantly influences the unmixing quality. For this image, there is a favorable regularity in dependence of sw: For the same number of clusters, increasing sw provides both lower ABU RMSE and EM RMSE. This means that the more emphasis is put on spectral similarity instead of spatial adjacency, the better the unmixing results.

Figure 5.17 shows the effect of the chosen distance measure and sw for UNCLI applied on the same image with 30 dB SNR. It can be stated from this and the other observed images (results not shown) that when noise is present, the ABU and EM RMSE curves are not shaped as regularly as in the noise-free case; however, a higher value of sw tends to provide better error values.

Regarding k-means clustering, the most important parameter apart from the number of clusters is the chosen distance measure. As mentioned

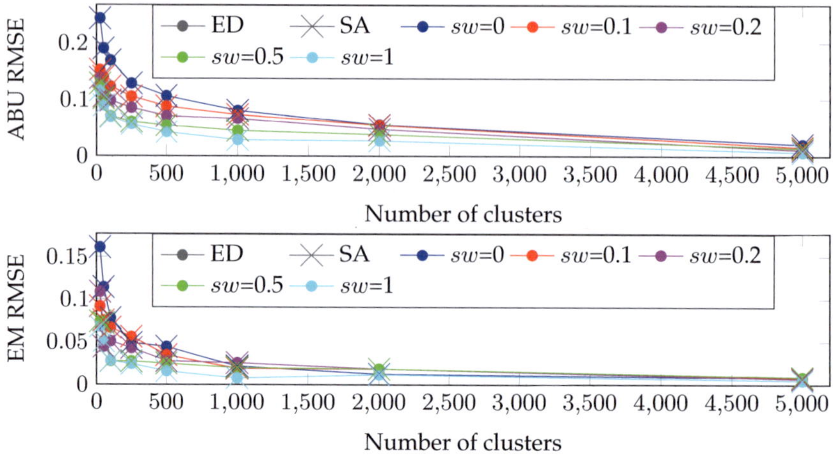

Figure 5.16 Urban Detail6, no noise, RHSEG clustering: dependence of the N-FINDR unmixing results on sw and the chosen distance measure. The data points of both distance measures ED and SA are very close and most often overlap.

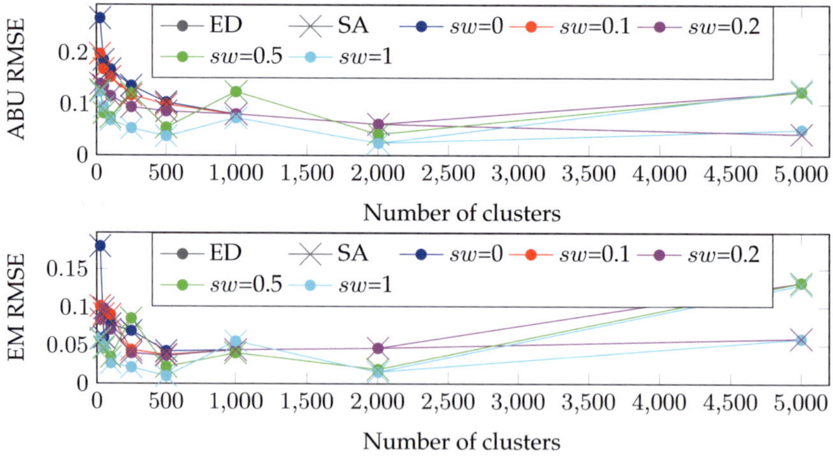

Figure 5.17 Urban Detail6 at 30 dB SNR, RHSEG clustering: dependence of the N-FINDR unmixing results on sw and the chosen distance measure.

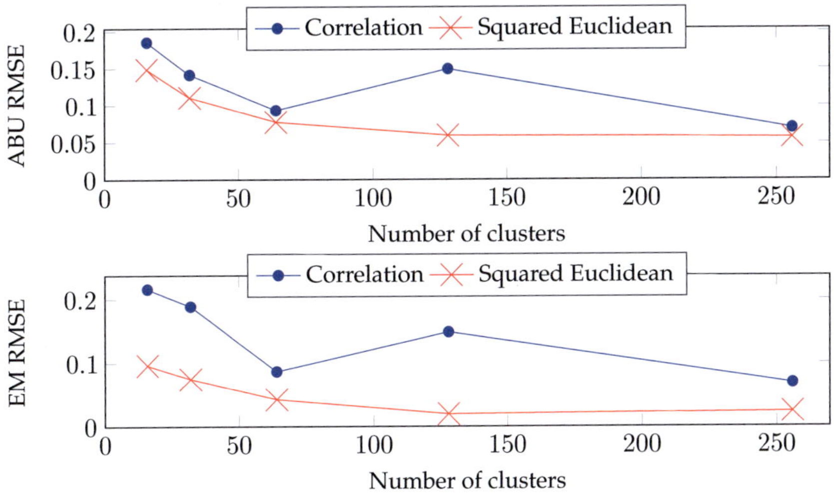

Figure 5.18 Urban Detail6 at 30 dB SNR, k-means clustering: dependence of the N-FINDR unmixing results on the chosen distance measure. The squared Euclidean distance provides lower error values.

before, Euclidean distance and correlation were used for the presented experiments. Figure 5.18 exemplary shows the results obtained with the Urban Detail6 image, SNR 30 dB. As can be seen, the Euclidean distance provides the lower abundance and endmember errors. This holds not only for this image, but also for the other considered images with only a few exceptions at isolated cluster numbers. Although the unmixing of k-means clusters exhibits a continuous decay for increasing cluster number (see Fig. 5.14 and Fig. 5.15), the drawback of this clustering method is its calculation time which grows quadratically with the number of clusters. This would mean that calculating twice as many clusters takes four times as long, see Fig. 5.19. For large cluster numbers, the application of UNCLI therefore does not provide a time saving, as the time required for clustering exceeds the saving in unmixing time.

The ISODATA clustering algorithm is very similar to k-means. Both start by defining randomly chosen pixel spectra to cluster centers. Iteratively, the remaining pixels are assigned to one of the clusters by deciding for the cluster the considered pixel has the least distance to and recalculating the

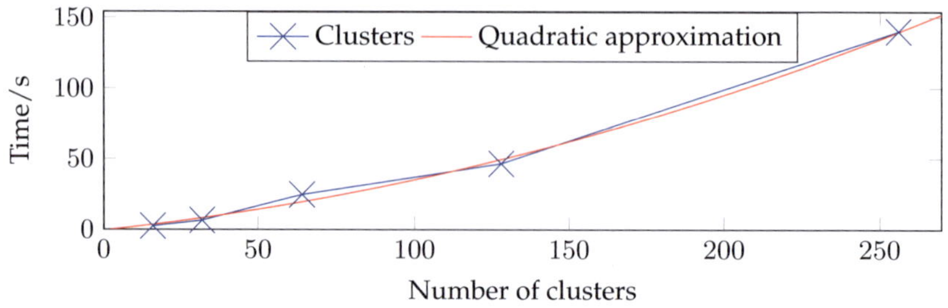

Figure 5.19 Urban Detail6 at 30 dB SNR, k-means clustering, squared Euclidean distance: clustering calculation time.

cluster means. The main difference compared to k-means is that ISODATA splits and merges clusters depending on various algorithm parameters such as, e.g., the distance between cluster centers or the cluster standard deviation. This results in ISODATA finding the number of output clusters on its own. Although the large number of parameters (maximum number of clusters, minimum number of elements in one cluster before it is eliminated, number of clusters that are mixed within an iteration, threshold for cluster splitting, threshold for cluster fusion, maximum number of iterations) allows for a sensitive tuning of the number of clusters and their size, this is also a main drawback of this method, as it can be very hard to find the best parameters. For the presented analyses, several different combinations of parameter values were considered, but no further parameter optimization was performed. With the chosen combinations, only low cluster numbers were realized (see Fig. 5.14 and Fig. 5.15). Larger cluster numbers would result in significantly higher time demand, similarly to k-means. Due to the complexity of the parameter selection, which results in more complicated dependence of the result on the chosen parameter values than in the case of, e.g., RHSEG (cf. Fig. 5.16), the ISODATA results will only be given in the following. No detailed discussion of the influence of the parameters will be conducted.

The main parameters influencing the SP clustering results are the minimum number of pixels in each cluster and the parameter c controlling the likelihood that clusters merge; the larger c, the higher the likelihood. For constant c, the influence of the parameters is shown in Fig. 5.20. Due to

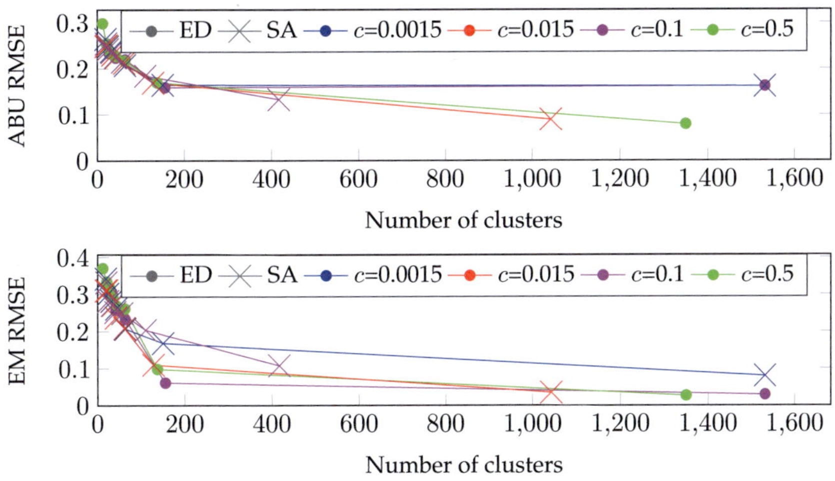

Figure 5.20 Urban Detail6 at 30 dB SNR, SP clustering: dependence of the N-FINDR unmixing results on the chosen distance measure and the algorithm parameter c. The data points for constant c belong to the minimum number of pixels in each cluster; this parameter was chosen to 2, 20, 40, 60, 80, 100, 120, where 2 belongs to the rightmost data point and 120 to the leftmost.

the fact that no clear influence of these parameters on the unmixing results can be identified, all parameter combinations will be incorporated in the following analyses.

After having analyzed the influence of the clustering method on the unmixing results, only selected parameters of the other clustering methods will be presented for clarity in the following: the RHSEG results will be shown for $sw = 0$ and $sw = 1$, considering only the ED as distance measure. ED is also for k-means the only distance measure that will be displayed. For ISODATA and SP, both ED and SA as well as all used parameter combinations will be presented.

5.4.4 Supervised Unmixing – UNCLI Run Time

After having discussed the influence of the various clustering methods and their parameters on the clustering results, it is interesting to turn to another

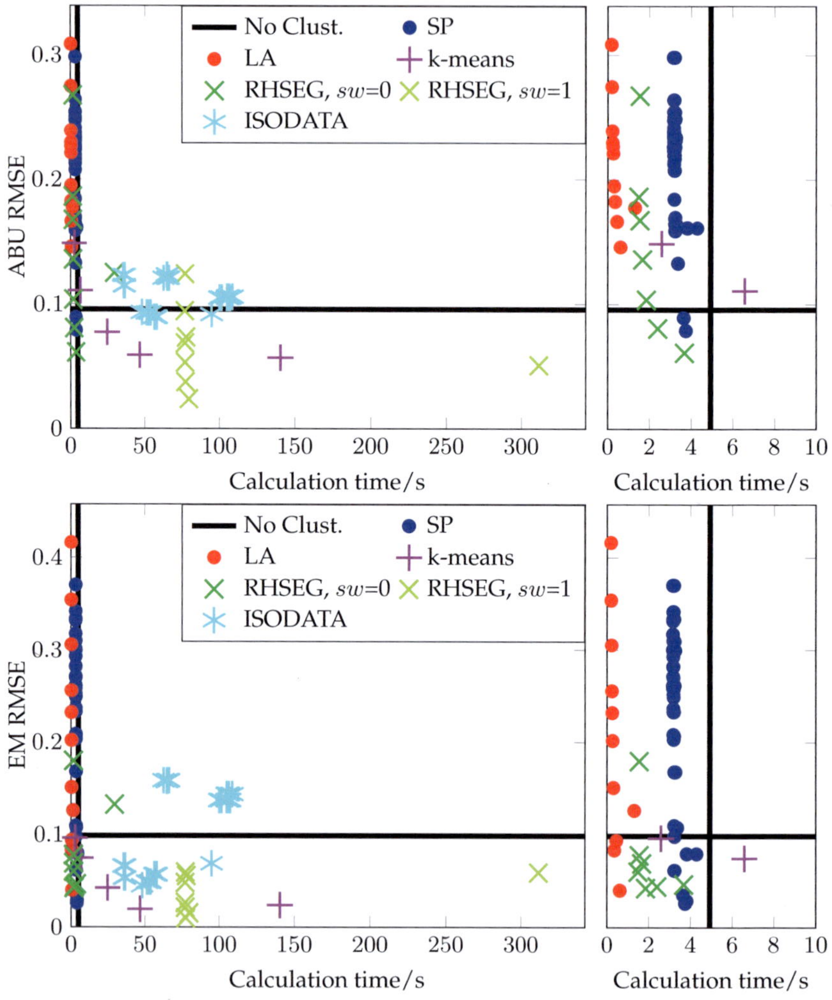

Figure 5.21 Urban Detail6, SNR 30 dB: unmixing with N-FINDR endmember initialization. The thick black lines represent the time and the error values obtained with the unclustered image. All results (left), enlarged section (right). Ideally, the clustering provides data points to the left of the vertical black line and below the horizontal one: better results are obtained in less time. For this image, only some RHSEG and some SP clusterings fulfill this condition for both abundances and endmembers.

interesting characteristic of the UNCLI method: its influence on the calculation time. All calculations have been performed on a desktop computer having an Intel i7-3770K processor and 16 GB RAM. The processor has 4 cores that can run 8 threads in parallel. No adaption of the respective method's implementation to the hardware architecture was performed, which means that these implementations as well as the actual unmixing (i.e., calculation of endmembers and abundances) may run on a different number of threads. The used RHSEG implementation, for instance, uses only one thread. However, the resulting computing times give a good idea of what qualitative time saving can be obtained by using UNCLI instead of unmixing the unclustered image.

In Fig. 5.21, the unmixing results of the Urban Detail6 image with 30 dB SNR and N-FINDR endmember initialization are shown. In contrast with Figs. 5.14 and 5.15, the abscissa this time does not show the number of clusters, but the calculation time. An additional vertical black line indicates the time required for unmixing the unclustered image. The ultimate goal of UNCLI is to provide better unmixing results in shorter time than unmixing of the clustered image does. This means that the UNCLI data points should be located in the lower left rectangle pictured by the horizontal and vertical lines of the unclustered images. If it is not possible to reach both goals, UNCLI should achieve smaller error values in more time (lower right rectangle) or suboptimal error values in shorter time (upper left rectangle). For the considered image, this is the case for almost all UNCLI results. The resulting best abundance maps of each method are displayed in Fig. 5.22.

After having analyzed the remotely sensed Urban Detail6 image with 30 dB SNR in detail, it is interesting to see how UNCLI performs on images acquired in the lab. By analogy with Fig. 5.21, Fig. 5.23 shows the unmixing results for the White Powders dataset. It can be seen that the clustering methods that are focusing on the spatial relation between the image pixels, such as LA, SP and RHSEG with sw=0, perform especially well. This is due to the spatial structure of the images, as the mixing ratio within each box compartment is constant, meaning that methods promoting spatial smoothness have the potential of considerably improving the unmixing results. Even LA, i.e., pure averaging of pixels within rectangular patches, performs better than straightforward unmixing of the original, unclustered image.

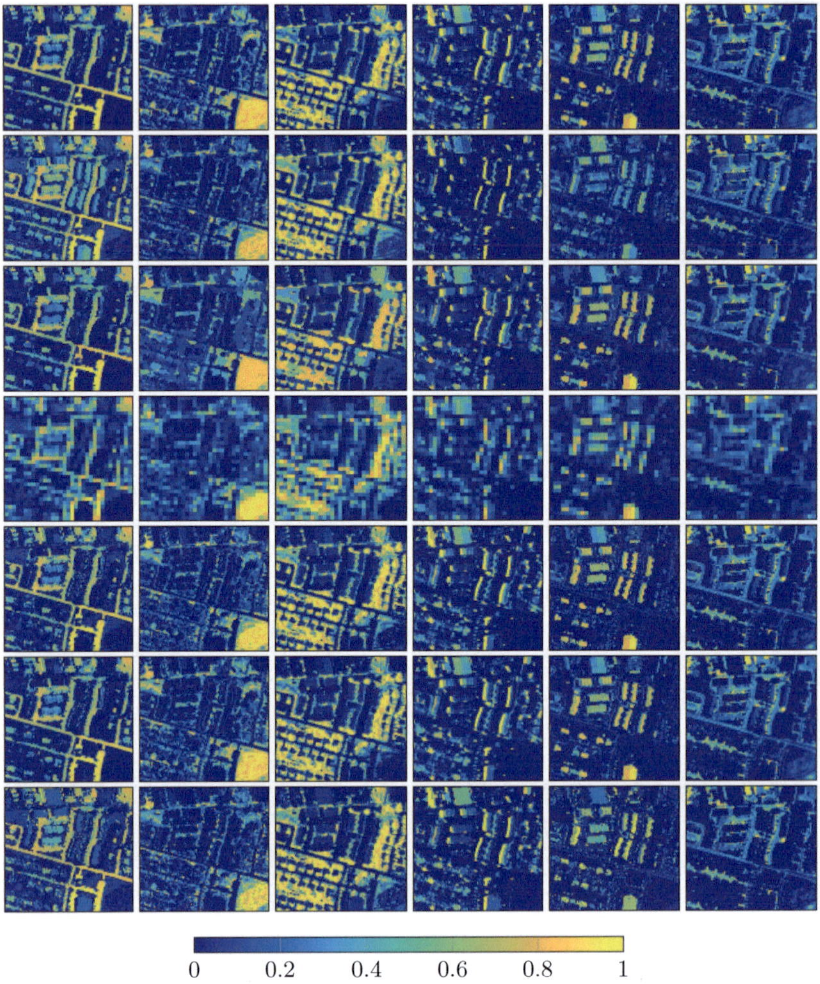

Figure 5.22 Urban Detail6, SNR 30 dB, N-FINDR endmember initialization: best abundance results of each clustering method. Row by row from top to bottom: ground truth; original, unclustered image; SP; LA; k-means; RHSEG; ISODATA. The unclustered image leads to a bad abundance estimation of the first and fifth endmember; SP, k-means and especially RHSEG lead to a considerable improvement.

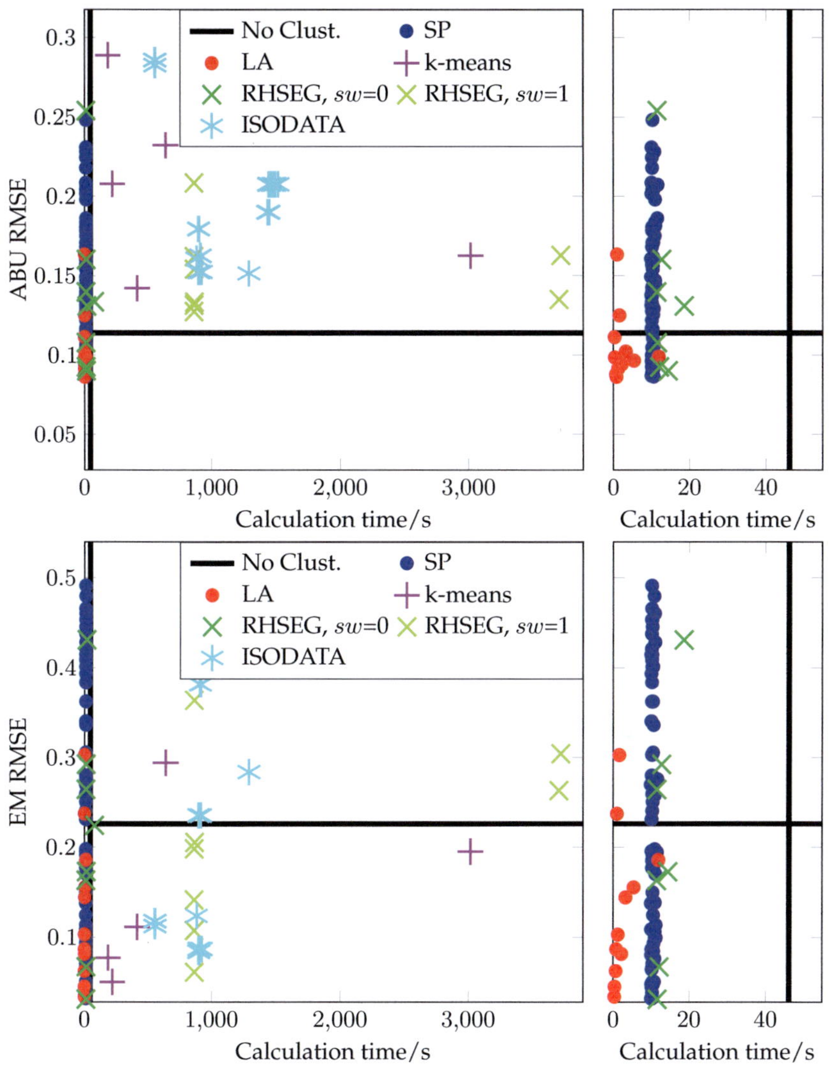

Figure 5.23 White Powders image: unmixing with N-FINDR endmember initialization. The thick black lines represent the time and the error values obtained with the unclustered image. All results (left), enlarged section (right). The clustering methods focusing on spatial neighborhood provide lower error values in much shorter time than full image unmixing.

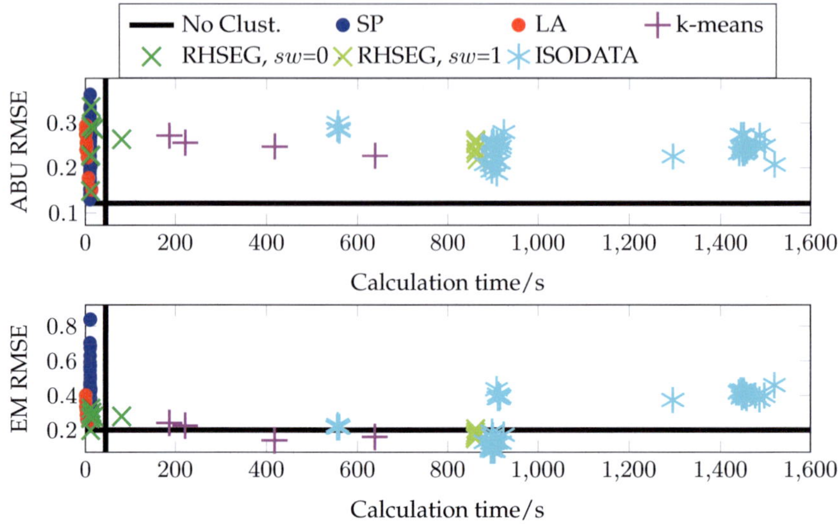

Figure 5.24 White Powders image: supervised unmixing with SISAL initialization with the same parameter τ for both the original and the clustered images. Two RHSEG (sw=1) and one k-means result requiring more than 1600 s are cut off for zooming and better visibility. For not adapted τ, UNCLI yields worse results than full image unmixing.

So far, only the results of the N-FINDR endmember initialization for supervised unmixing have been discussed. SISAL endmember initialization also has been investigated, but it requires the tuning of the parameter τ, which depends on the size of the input image. Due to the fact that the clustered images are smaller, τ should be adapted to the size of each clustered image. While Fig. 5.24 shows the unmixing results with τ of each clustered image chosen to the one of the original image, Fig. 5.25 presents the results when τ is adapted for each clustered image individually. This individual adaption leads to a better endmember estimation, which in the second step provides a better abundance estimation.

Let us now consider all used images described in Sec. 5.1.1, i.e., both the remotely sensed and the laboratory ones. Table 5.7 shows the unmixing results for original and clustered images. The initialization was calculated by N-FINDR. For each clustering method, the parameter combination that

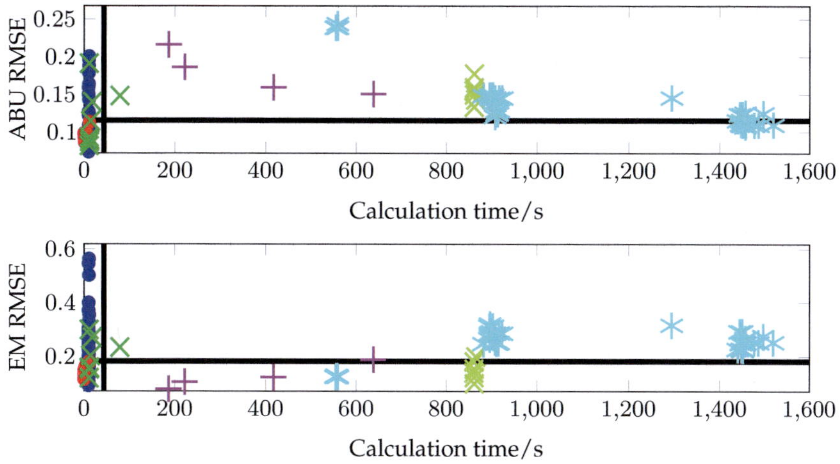

Figure 5.25 White Powders image: supervised unmixing with SISAL initialization with τ adapted to each clustered image. This provides better results than fixed τ, cf. Fig. 5.24

provides the lowest ABU RMSE is selected. The ABU RMSE values are given in the first horizontal block, the corresponding EM RSME values in the second block, its calculation time in the third block and the cluster number in the last one. The results of the clustered images are marked green if they are better than the results of the unclustered image and red if not. If they are marked red, they are in many cases only slightly worse. Shorter calculation times than full image unmixing are highlighted in green, longer in red. For most images and clustering methods, UNCLI yields better endmember estimations. Considering the abundance estimation, UNCLI provides lower error values in the presence of noise.

Comparing all methods, RHSEG with sw=1 in most cases provides better unmixing results in terms of both ABU and EM RMSE than unmixing of the full image. If it provides higher error values, the difference is small. The calculation time for RHSEG with sw=1, however, is much longer because of the time-consuming clustering. SP and LA run faster on all images; although the abundance quality is not as good as the one of full image unmixing, it is in most cases only some percent worse.

Table 5.7 N-FINDR initialization: best UNCLI results, rated by ABU RMSE. RHSEG with sw=0 is abbreviated R0, with sw=1 correspondingly. k-means is k for short and I denotes ISODATA. Many UNCLI results marked in red have just slightly higher error values than the unclustered image. Blocks from top to bottom: ABU RMSE, EM RMSE, calculation time/s, number of clusters.

Clustering	Jasper	Jasper 20 dB	Jasper 30 dB	Urban Detail4	Urban Detail4 20 dB	Urban Detail4 30 dB	Urban Detail6	Urban Detail6 20 dB	Urban Detail6 30 dB	White Powders	Spatial Mixing 10	Spatial Mixing 20	Spatial Mixing 30
no	0.000	0.030	0.105	0.011	0.213	0.014	0.009	0.181	0.096	0.114	0.140	0.147	0.100
SP	0.044	0.119	0.054	0.059	0.147	0.067	0.079	0.142	0.079	0.087	0.154	0.150	0.130
LA	0.080	0.081	0.080	0.113	0.148	0.113	0.112	0.119	0.146	0.086	0.171	0.170	0.162
k	0.014	0.025	0.017	0.024	0.039	0.032	0.056	0.066	0.057	0.142	0.141	0.125	0.100
R0	0.005	0.043	0.010	0.015	0.048	0.021	0.021	0.125	0.061	0.090	0.141	0.135	0.114
R1	0.000	0.024	0.008	0.011	0.035	0.016	0.007	0.054	0.024	0.127	0.140	0.122	0.119
I	0.035	0.036	0.030	0.069	0.067	0.047	0.088	0.087	0.089	0.151	0.135	0.141	0.122
no	0.000	0.124	0.120	0.012	0.327	0.043	0.006	0.271	0.099	0.226	0.694	0.722	0.375
SP	0.002	0.055	0.015	0.019	0.093	0.028	0.018	0.072	0.026	0.085	0.199	0.363	0.320
LA	0.004	0.062	0.024	0.021	0.049	0.035	0.011	0.069	0.040	0.087	0.289	0.337	0.700
k	0.004	0.036	0.008	0.018	0.044	0.026	0.023	0.036	0.024	0.111	0.614	0.449	0.293
R0	0.000	0.100	0.040	0.012	0.087	0.049	0.008	0.146	0.046	0.173	0.515	0.361	0.224
R1	0.000	0.060	0.038	0.012	0.041	0.038	0.005	0.035	0.015	0.141	0.581	0.352	0.425
I	0.014	0.024	0.009	0.026	0.034	0.026	0.072	0.055	0.057	0.284	0.382	0.451	0.334
no	6.8	3.6	4.1	4.1	3.7	3.7	7.1	4.6	4.9	46.3	33.1	15.3	11.3
SP	3.8	3.5	3.7	3.7	3.7	3.5	3.7	3.8	3.7	10.5	6.3	5	4.9
LA	1.6	1	1	1.1	0.5	0.9	1.2	1.2	0.6	0.8	3.4	4.6	2.8
k	123.3	347	245.3	105.8	249.5	168.6	111.9	127.1	140.3	417.4	550.1	460.8	296.4
R0	47.3	3.7	47.3	29.8	3.4	27.8	29.9	3.6	3.7	14.4	119.3	5.3	3.7
R1	217.5	137.9	469.7	198	105	349.9	242.7	95.4	79.8	861.4	763.3	187	86
I	67.9	69.7	72.3	65.1	52.7	103.7	115.8	58	57.5	1295.4	428.6	220.1	150.7
SP	1132	121	775	1421	1531	1061	1360	1531	1350	719	946	900	1531
LA	2500	2500	2500	2500	1089	2500	2500	2500	1089	1512	6566	7400	4900
k	256	256	256	256	256	256	256	128	256	64	256	256	256
R0	5000	2000	5000	5000	2000	5000	5000	2000	2000	1000	5000	1000	1000
R1	5000	1000	5000	5000	500	5000	5000	500	2000	100	250	250	1000
I	62	61	72	76	61	71	76	67	66	54	71	69	72

5.4.5 Unsupervised Unmixing Results

The results discussed in the previous sections have been obtained by supervised unmixing, i.e., determining the endmembers with an EEA first and using the resulting endmembers for calculating the abundances afterwards. For the unsupervised unmixing discussed in this section, the used methods calculate both endmembers $\mathbf{M}$ and abundances $\mathbf{A}$ alternatingly and at the end provide a joint estimation of both matrices. MVC and MiniDisCo (see Sec. 3.2.2) are used as unmixing methods. The MVC endmember matrices are initialized with randomly selected pixels of the full image, not from the clustered images, to allow for a fair comparison. The initial endmember matrix for MiniDisCo is all zero, as directed by the method description in [63]. The abundance matrix is initialized with the constant value $1/R$. The respective regularization parameter μ_{spec} was adapted to the size of the (clustered) images: $\mu_{\text{spec,clust}} = C/I \cdot \mu_{\text{spec}}$, where μ_{spec} is the optimal value for the full image and $\mu_{\text{spec,clust}}$ the one for the respective clustered image. For both unmixing methods, the number of iterations was set to $T = 500$. Each iteration consists of 5 PQN abundance updates and 5 PQN endmember updates. Using the same pattern as Tab. 5.7, Tab. 5.8 shows the MVC results. While UNCLI in most cases performs considerably faster than full image unmixing (depending on image and clustering method up to 30 times), it provides error values that are a bit worse for most images and clustering methods. However, MVC provides for many images very good unmixing results which are hard to beat; still, UNCLI provides especially for k-means and RHSEG high quality unmixing results for both endmember and abundances.

For the MiniDisCo results shown in Tab. 5.9, the conclusion is the same: abundance and endmember error values are in many cases only slightly worse than the ones of full image unmixing, but there is considerable time saving.

5.4.6 Summary

In this section, the application of UNCLI has been investigated. Several clustering algorithms have been applied to some remotely sensed hyperspectral images as well as to some images acquired under laboratory conditions. UNCLI yields in the case of both supervised and unsupervised

Table 5.8 UNCLI results obtained using MVC, ordered by best ABU RMSE. Blocks from top to bottom: ABU RMSE, EM RMSE, calculation time/s, number of clusters.

Clustering	Jasper	Jasper 20 dB	Jasper 30 dB	Urban Detail4	Urban Detail4 20 dB	Urban Detail4 30 dB	Urban Detail6	Urban Detail6 20 dB	Urban Detail6 30 dB	White Powders	Spatial Mixing 10	Spatial Mixing 20	Spatial Mixing 30
no	0.007	0.021	0.009	0.011	0.022	0.013	0.039	0.083	0.053	0.169	0.173	0.195	0.162
SP	0.045	0.117	0.056	0.061	0.146	0.070	0.088	0.152	0.100	0.142	0.180	0.190	0.174
LA	0.082	0.083	0.082	0.107	0.108	0.107	0.121	0.134	0.127	0.156	0.191	0.224	0.179
k	0.016	0.026	0.019	0.026	0.033	0.027	0.063	0.087	0.076	0.172	0.173	0.185	0.154
R0	0.011	0.022	0.012	0.018	0.026	0.018	0.042	0.083	0.059	0.137	0.185	0.184	0.165
R1	0.008	0.020	0.009	0.016	0.025	0.014	0.038	0.083	0.056	0.170	0.179	0.187	0.161
I	0.038	0.037	0.031	0.054	0.049	0.044	0.097	0.104	0.094	0.154	0.148	0.179	0.155
no	0.011	0.020	0.012	0.013	0.015	0.013	0.057	0.192	0.152	0.485	1.023	1.006	1.128
SP	0.017	0.045	0.023	0.025	0.034	0.028	0.054	0.194	0.199	0.435	0.862	1.251	1.098
LA	0.021	0.025	0.021	0.019	0.019	0.019	0.071	0.201	0.179	0.463	0.931	1.116	1.131
k	0.016	0.019	0.014	0.028	0.017	0.021	0.091	0.223	0.210	0.505	0.982	1.337	1.101
R0	0.017	0.023	0.015	0.023	0.017	0.022	0.071	0.217	0.189	0.484	0.951	1.093	1.124
R1	0.015	0.018	0.011	0.027	0.015	0.015	0.088	0.196	0.176	0.434	0.903	1.331	1.081
I	0.030	0.022	0.016	0.021	0.023	0.024	0.096	0.222	0.134	0.510	0.875	1.370	1.125
no	185.7	187.1	117.2	154.3	92.8	95.1	269.9	269.5	270.9	3997.8	2041.4	1000.3	687.8
SP	28.5	15.4	18.9	32.8	24.7	18.3	55.0	58.8	54.7	79.9	70.2	31.8	69.0
LA	57.8	51.5	39.4	47.9	27.9	31.0	80.8	83.5	82.3	992.6	463.5	266.6	167.2
k	132.5	354.9	251.2	116.0	255.0	174.0	128.5	222.9	157.6	656.5	571.7	163.8	32.8
R0	132.0	127.8	110.3	108.2	86.5	78.8	175.6	169.5	174.9	42.8	40.4	93.4	231.8
R1	294.0	617.8	529.2	286.0	467.4	405.7	389.4	167.5	461.6	876.8	787.3	207.0	101.1
I	78.2	69.5	77.0	71.3	57.4	59.8	121.6	67.9	106.6	1535.5	441.0	238.3	149.3
SP	1132	121	775	1421	1531	1061	1360	1531	1350	1471	1531	444	1531
LA	2500	2500	2500	2500	2500	2500	2500	2500	2500	30400	14800	7400	4900
k	256	256	256	256	256	256	256	256	256	128	256	128	32
R0	5000	5000	5000	5000	5000	5000	5000	5000	5000	500	250	2000	5000
R1	5000	5000	5000	5000	5000	5000	5000	2000	5000	50	250	50	50
I	62	61	72	77	61	68	77	67	56	65	73	69	53

Table 5.9 UNCLI results obtained using MiniDisCo, ordered by best ABU RMSE. Blocks from top to bottom: ABU RMSE, EM RMSE, calculation time/s, number of clusters. Note that many UNCLI abundance errors which are higher than the ones of full image unmixing are less than one percent worse.

Clustering	Jasper	Jasper 20 dB	Jasper 30 dB	Urban Detail4	Urban Detail4 20 dB	Urban Detail4 30 dB	Urban Detail6	Urban Detail6 20 dB	Urban Detail6 30 dB	White Powders	Spatial Mixing 10	Spatial Mixing 20	Spatial Mixing 30
no	0.046	0.055	0.049	0.062	0.066	0.062	0.095	0.103	0.102	0.207	0.193	0.158	0.141
SP	0.070	0.129	0.077	0.094	0.161	0.099	0.126	0.171	0.129	0.113	0.180	0.164	0.148
LA	0.097	0.099	0.099	0.130	0.130	0.129	0.157	0.157	0.154	0.171	0.198	0.189	0.168
k	0.054	0.056	0.054	0.072	0.071	0.067	0.110	0.116	0.122	0.135	0.157	0.146	0.126
R0	0.047	0.053	0.051	0.059	0.067	0.066	0.100	0.110	0.103	0.104	0.171	0.140	0.133
R1	0.048	0.052	0.050	0.064	0.067	0.064	0.108	0.109	0.097	0.160	0.171	0.141	0.126
I	0.062	0.061	0.060	0.079	0.068	0.067	0.126	0.120	0.120	0.133	0.145	0.139	0.129
no	0.069	0.085	0.074	0.066	0.066	0.066	0.171	0.144	0.184	0.809	1.187	1.197	1.082
SP	0.081	0.092	0.082	0.081	0.092	0.083	0.141	0.181	0.175	0.371	0.823	0.867	0.760
LA	0.071	0.079	0.077	0.079	0.078	0.076	0.212	0.201	0.159	0.528	1.189	1.095	0.992
k	0.084	0.075	0.077	0.079	0.069	0.071	0.146	0.181	0.230	0.320	1.121	0.642	0.944
R0	0.073	0.081	0.080	0.066	0.067	0.072	0.173	0.173	0.183	0.214	1.212	0.955	0.928
R1	0.076	0.078	0.075	0.069	0.064	0.068	0.143	0.162	0.155	0.532	1.264	0.533	0.739
I	0.077	0.075	0.075	0.073	0.055	0.062	0.183	0.126	0.129	0.327	1.082	0.764	0.669
no	220.0	244.4	169.9	166.2	140.8	157.0	269.2	266.5	270.2	4114.6	2079.2	1069.6	700.4
SP	50.4	17.8	23.6	35.6	29.1	20.2	54.5	58.1	47.0	28.4	45.5	33.9	27.6
LA	79.0	65.3	50.3	47.6	37.7	35.0	82.6	78.2	80.6	123.0	467.6	133.7	177.2
k	141.4	360.1	255.2	118.9	261.8	177.0	59.0	142.1	156.7	437.1	129.3	42.5	138.3
R0	143.7	157.7	153.3	91.7	99.8	98.8	177.8	172.5	157.1	28.0	69.5	87.0	241.4
R1	293.7	624.7	570.4	63.9	113.0	427.8	82.0	120.7	461.5	3889.3	818.6	202.7	109.5
I	84.9	74.3	84.9	120.1	61.9	63.2	131.0	71.7	71.0	913.4	448.5	239.9	162.6
SP	1132	121	775	1421	1531	1061	1531	1531	1350	44	644	443	306
LA	2500	2500	2500	2500	2500	2500	2500	2500	2500	3339	14800	3283	4900
k	256	256	256	256	256	256	128	128	256	64	64	32	128
R0	5000	5000	5000	5000	5000	5000	5000	5000	5000	50	1000	2000	5000
R1	5000	5000	5000	250	250	5000	250	500	5000	5000	1000	25	250
I	62	61	72	76	61	70	76	67	66	73	73	69	68

unmixing for most images and noise levels lower endmember estimation errors than full image unmixing. Regarding the abundance estimation quality, especially supervised unmixing (endmember estimation provided by N-FINDR) can be greatly improved by UNCLI. The unsupervised unmixing algorithms MiniDisCo and especially MVC provide very good abundance results on the full images. For this reason, it is hard for UNCLI to yield lower error values. In many cases, though, the difference is a few per cent, and UNCLI provides the results faster than full image unmixing.

From the presented analyses, RHSEG can be recommended as the method of choice for UNCLI. The main reason for this is the easy parameter selection. Unlike ISODATA, for instance, which has lots of parameters whose impact on the unmixing results is not obvious, RHSEG has only two major parameters. The number of clusters can be chosen beforehand; it seems to be a good choice to allow for roughly 5 to 10 pixels per cluster. Apart from the number of clusters, the parameter sw allows for a tradeoff between unmixing quality and calculation time: the larger sw, the more accurate the results, but the longer the calculation time.

UNCLI with SP and LA calculates particularly fast; the unmixing can be accelerated several times compared to full image unmixing. For the investigated images, the relative time saving becomes even greater with increasing spatial image size. One natural extension of UNCLI is therefore the application of a final step (FS). After the UNCLI calculations, the endmembers provided by UNCLI are kept fixed, and the full resolution abundances are refined using these fixed endmembers. This can be done quickly in one least-squares step. This approach makes use of the benefit of fast endmember and relatively coarse abundance calculation. Due to UNCLI's more accurate endmember estimation in the presence of noise, it is likely that the final step provides even better abundance values than full image unmixing. The time saving is particularly large for LA and SP, see Tables 5.8 and 5.9. Figure 5.26 exemplarily shows the results of the application of the final step after unsupervised MVC. Note that the abundance errors of LA and RHSEG with $sw = 0$ are greatly improved; this makes it possible to combine the fast calculation in Tab. 5.8 with satisfying abundance accuracy.

The combination of LA UNCLI with the final step is very similar to PYRU (see Sec. 4.3). LA divides the image in patches and estimates abun-

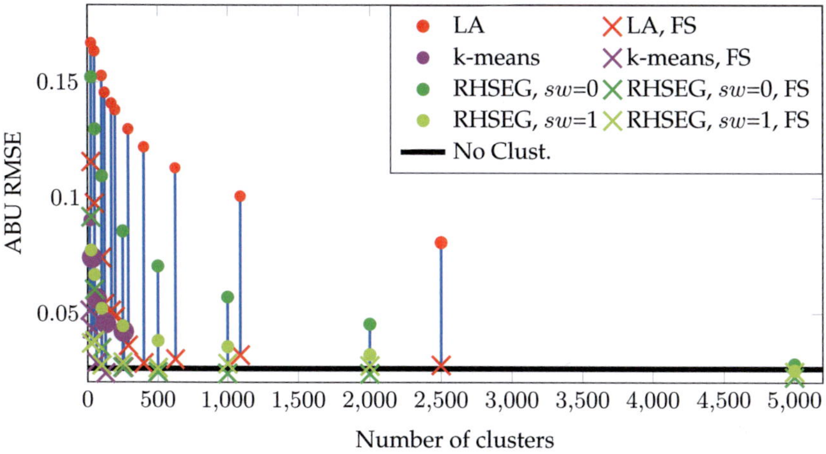

Figure 5.26 Urban Detail6, no noise, UNCLI with MVC: improvement after recalculating the full resolution abundances in a final step (FS).

dances for these patches. This is the same principle as using the image pyramid, for which patches of the full resolution image are used for calculating pixels of the higher pyramid level. The only difference is that in the final UNCLI step, only the abundances are recalculated on the full resolution level and the endmembers not. PYRU refines both endmember and abundances. This similarity leads over to the PYRU results presented and discussed in the next section. Due to the fact that the endmembers are recalculated on the ground pyramid level, it is likely that PYRU provides similar results as full image unmixing. Its goal is therefore to obtain the same unmixing quality as full image unmixing, but in less time.

5.5 PYRU

This section presents the results of the PYRU (pyramid unmixing) method from Sec. 4.3 combined with MVC and MiniDisCo spectral regularization (Sec. 3.2.2). As for all experiments presented in this thesis, PQN is used for the optimization of the unmixing objective functions, it will also be used for PYRU. Note that this is in contrast to [171], where the Armijo rule explained in Sec. 4.6 was used as optimization technique because of its advantages over fixed step sizes discussed in Sec. 4.6.

The number T of iterations on each pyramid level is determined by the formula (4.25), which means that the basis B and the prefactor V are the parameters defining T. For each iteration, $P = 5$ PQN abundance estimation steps and $M = 5$ endmember iteration steps are conducted. This is due to the fact that PQN selects the step size based on several past iterations and not the very last one, see Sec. 4.6.1, so it is reasonable to perform more than one step for each matrix. For all experiments, only the first image pyramid stage has been used, but with different reduction factors $N \in \{2, 4, 8, 16\}$. Choosing for example $N = 4$ is equivalent to using the original resolution level plus two reduced pyramid levels and skipping the calculation on the two levels in between.

5.5.1 PYRU with Minimum Volume Constraint (MVC)

PYRU is combined with the MVC method described in Sec. 3.2.2 and applied to all images introduced in Sec. 5.1.1. PYRU's main goal is to accelerate the unmixing process, i.e., the same objective function values should be obtained in shorter time than from the application of the respective plain unmixing method on the full resolution image. However, the acceleration should not come at the cost of abundance or endmember degradation. The values of objective function as well as ABU and EM RMSE over time for the Urban Detail4 image (no noise) are depicted in Fig. 5.27. Depending on the parameter choice, the time saving is about 50 %, meaning that the same value of the respective measure is reached about twice as fast as in the case of full image unmixing. In general, it can be said that the greater the prefactor V, the better the unmixing quality. The parameter V determines the number of iterations performed on the full resolution image; for the full image, B does not influence the number of iterations, as $k = 0$ in (4.25). From Fig. 5.27, it can be deduced that $N \in \{2, 4, 8\}$ are able to come to the same error values as full image unmixing, while $N = 16$ leads to slightly worse values. It is less suitable for this image; probably because important spatial information is lost on the first pyramid level, causing the optimization on the ground level to get stuck in a local minimum.

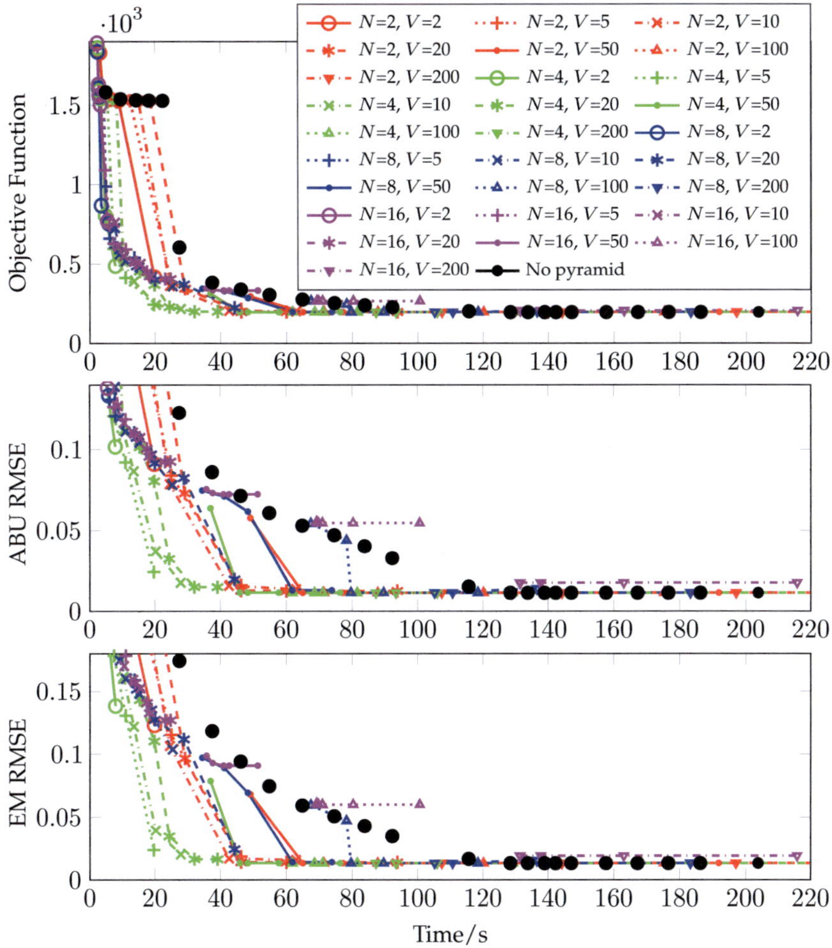

Figure 5.27 Urban Detail4 image without noise, PYRU with MVC: error values over time. Each connected line of data points is composed from data points having the same value of the prefactor V, but different bases B. The higher B, the more calculation time is required, because more iterations are performed on the first pyramid level $\mathbf{Y}^1$ in this case. The number of iterations on the full resolution is V, regardless of B (see Eq. (4.25)).

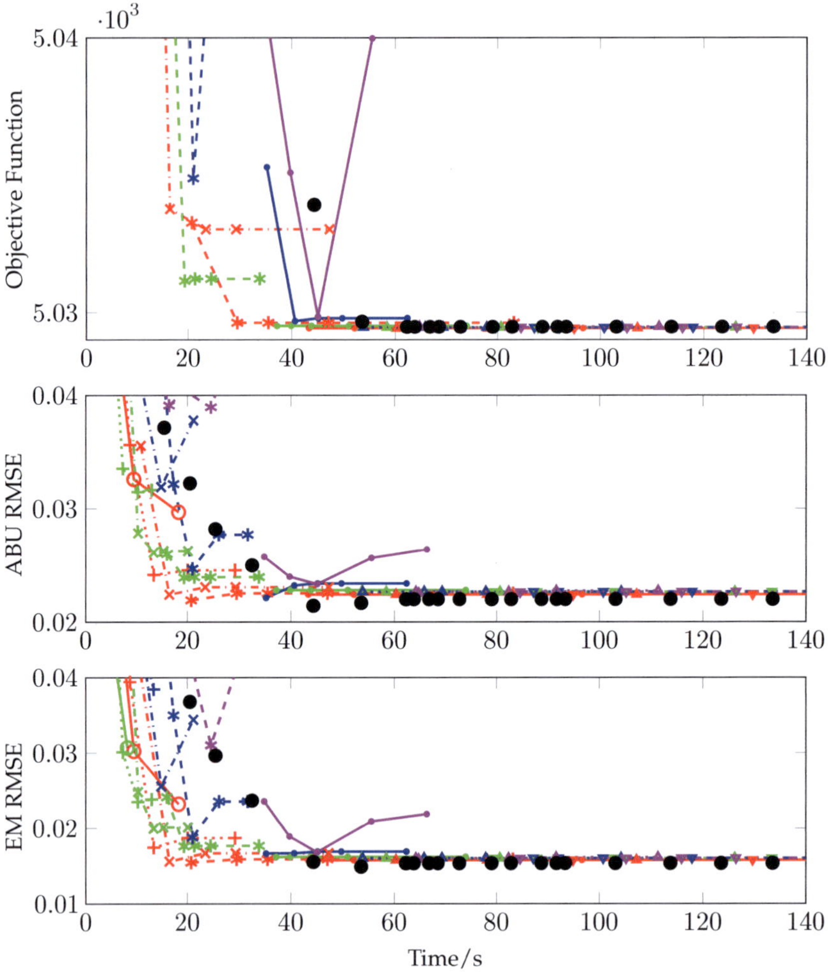

Figure 5.28 Urban Detail4 image, 20 dB SNR, PYRU with MVC: detailed view of the error values over time. Each connected line of data points is composed from data points having the same value of the prefactor V, but different bases B. The legend is the same as in Fig. 5.27.

The results for the same image with 20 dB SNR are illustrated in Fig. 5.28. This time, only a detail is shown. While the application of PYRU for sufficient high B and V leads to a slightly lower objective function value than unmixing of the full image, the ABU and EM RMSE values are slightly worse. However, note that the difference is well below one percent. Selecting $N = 2$ leads to a calculation time reduction of about 50 %, similar to the noise-free image. The parameter combination $N = 16$, $V = 50$ shows a strange behavior; $N = 16$ was already causing problems with the noise-free image.

A noise level of 30 dB leads to structurally very similar results as the ones for 20 dB shown in Fig. 5.28 regarding time requirement for PYRU and full image unmixing. For this reason, no plots are given. Only the ABU and EM RMSE values are a bit lower.

Analyzing the Urban Detail6 image leads to an even larger time saving with PYRU. The results for a noise level of 30 dB are shown in Fig. 5.29; the plots of different noise levels exhibit a similar structure. For the given noise level, the time saving is about a factor of six for $N = 4$ and a factor of four for $N = 2$. $N = 8$ and $N = 16$ do not provide acceptable results.

Applying PYRU with MVC to the Jasper image leads to time savings of about 10 to 50 %, depending on the parameter selection and the noise level.

The White Powders image is an image acquired in a laboratory environment, not remotely sensed, and compared to the images analyzed so far of very large size. For large images, PYRU can lead to great time savings; for the considered image, it performs about five times faster. The results are illustrated in Fig. 5.30. The endmember estimation quality does not improve steadily with time in the case of full image unmixing. This effect is also mirrored in the PYRU results.

The Spatial Mixing images are hard to unmix without taking endmember variability (Sec. 2.1.2) into account. This effect can also be observed in the results provided by the application of PYRU. The error plots given in Fig. 5.31 do not follow a continuous decay such as the ones of the other images.

It can be stated that PYRU with MVC performs very well and can, especially for larger images, lead to great time savings. This only holds for images for which the original MVC method applied to the full resolution image also performs well. If this is not the case, PYRU naturally

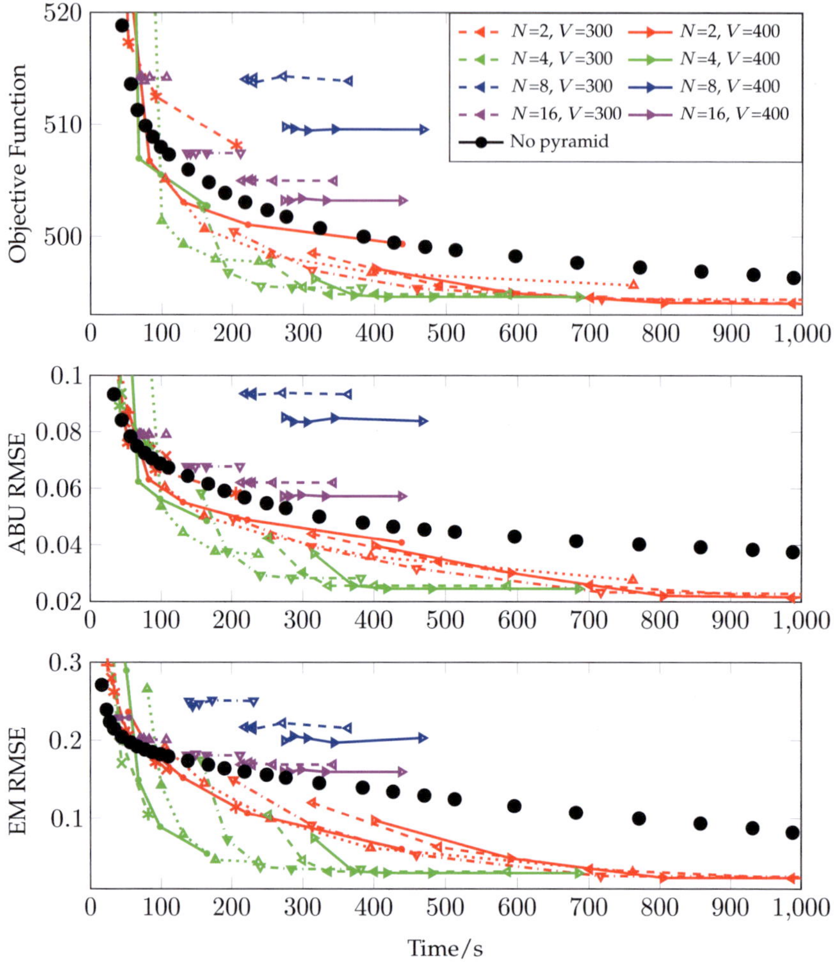

Figure 5.29 Urban Detail6 image, 30 dB SNR, PYRU with MVC: error values over time. Each connected line of data points is composed from data points having the same value of V, but different bases B. While $N = 2, 4$ lead to very good results, higher values are not feasible. See Fig. 5.27 for the legend; the legend of the present figure only shows the legend entries of additional curves for higher V.

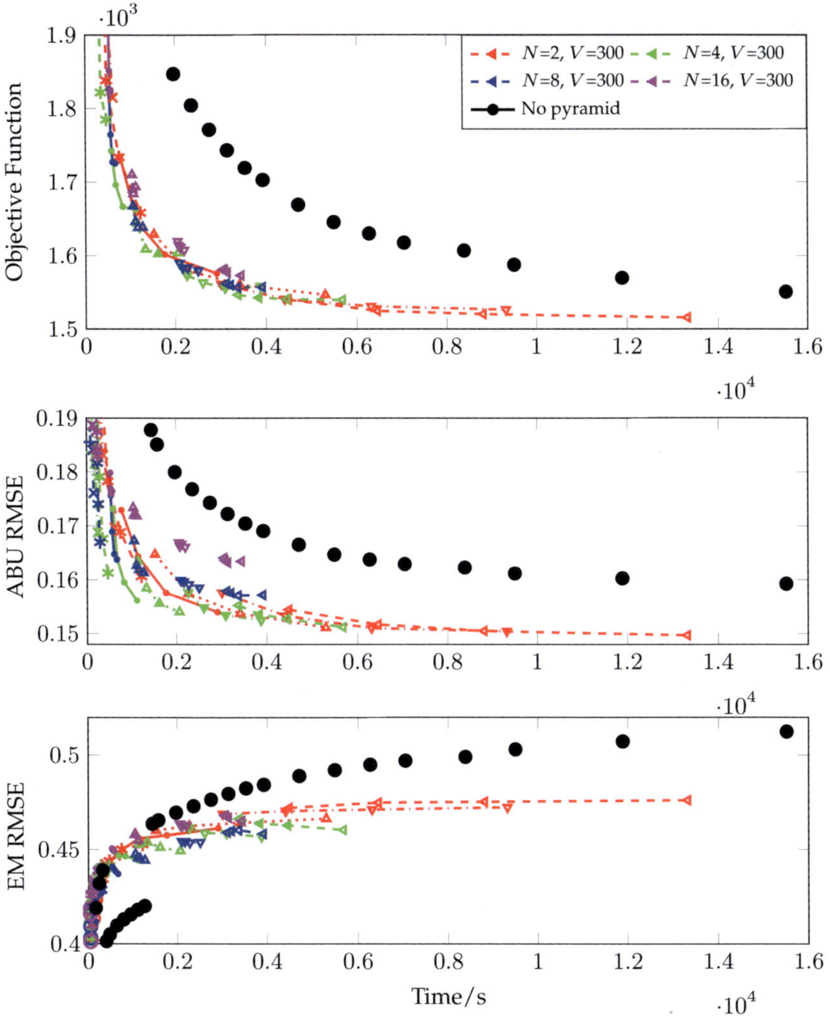

Figure 5.30 White Powders, PYRU with MVC: error values over time. Each connected line of data points is composed from data points having the same value of the prefactor V, but different bases B. PYRU with $N = 2$ runs about five times faster than full image unmixing. Especially for such large images, PYRU leads to an enormous absolute time saving. Apart for the data points with $V = 300$, the legend is the same as in Fig. 5.27.

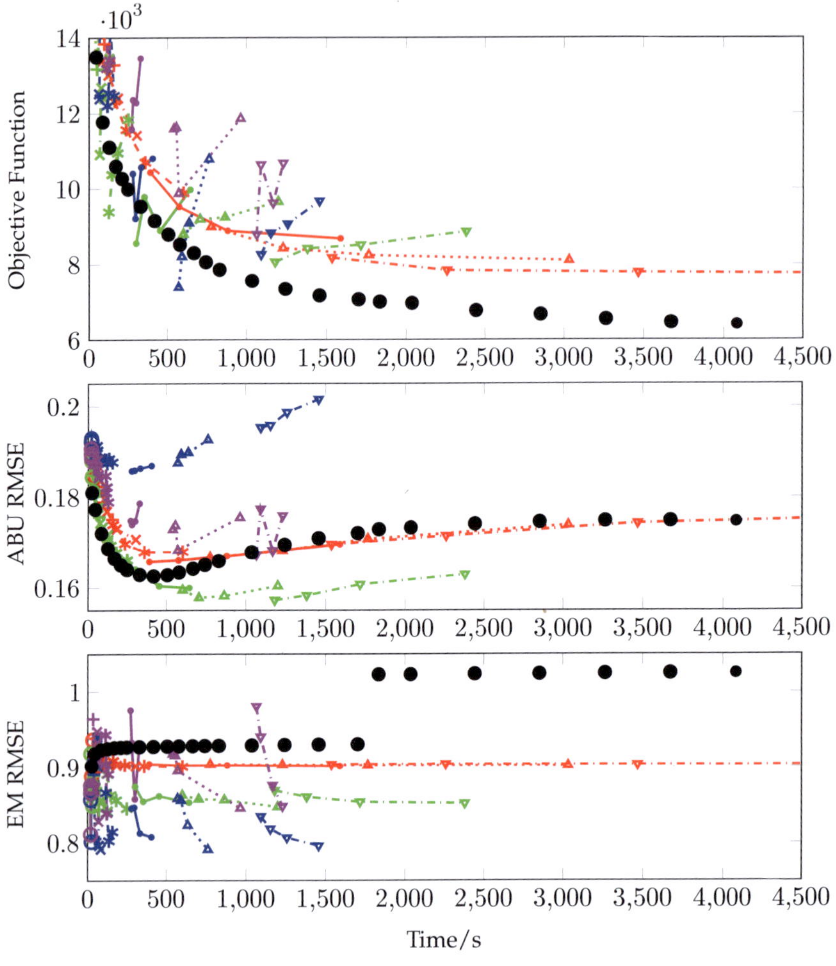

Figure 5.31 Spatial Mixing 10, PYRU with MVC: error values over time. Each connected line of data points is composed from data points having the same value of the prefactor V, but different bases B. Even the curves of full image unmixing do not decay steadily; the same holds for the error values of PYRU. The legend is the same as in Fig. 5.27.

cannot provide better results, as it is not an unmixing improvement, but acceleration method.

5.5.2 PYRU with Minimum Dispersion Constraint (MiniDisCo)

The combination of MiniDisCo with PYRU also has been implemented and its effect analyzed. By contrast to MVC, the endmember initialization does not consist of randomly selected image pixels, but the endmember matrix $\mathbf{M}$ is set to all zeros. PYRU in general passes the results calculated on one pyramid stage on to the following stage, i.e., the matrices $\mathbf{A}$ and $\mathbf{M}$ calculated on the first unmixing level are used as initial matrices for the subsequent level. This way, calculation effort is saved, especially because of the large reduction of the size of $\mathbf{A}$. For MiniDisCo, the question is if the endmember matrix should also be passed, or if the subsequent initialization should be a matrix of zeros again. From the analyses performed for this thesis, it is known that random initialization of the MiniDisCo endmember matrix can very easily lead to the algorithm getting stuck in local minima. For this reason, both ways of initializing $\mathbf{M}$ on the next level are considered: initialization with the result of the previous stage, or a matrix of zeros.

Figure 5.32 shows the error values for the Jasper image with 30 dB SNR. The estimated endmembers are assigned from one pyramid level to the next. For $N = 2$, the results are more or less comparable with the PYRU MVC results, but for all other values of N, especially the endmember estimation fails. When reinitializing the matrix $\mathbf{M}$ to zeros on each level (plots omitted), the endmember results become even worse; the ABU RMSE and objective function plots are similar to the ones in Fig. 5.32.

Unfortunately, PYRU with MiniDisCo does not provide the same results as pure MiniDisCo in shorter time; the unmixing quality becomes even worse. One possible explanation for this behavior is the promotion of flat endmembers by the MiniDisCo regularizer which minimizes the endmember variance. The higher pyramid level naturally contains flatter pixel spectra due to the averaging of neighboring pixels. It is not unlikely that this causes the optimization algorithm to get caught in a local minimum. Additionally, the original MiniDisCo procedure initializes the

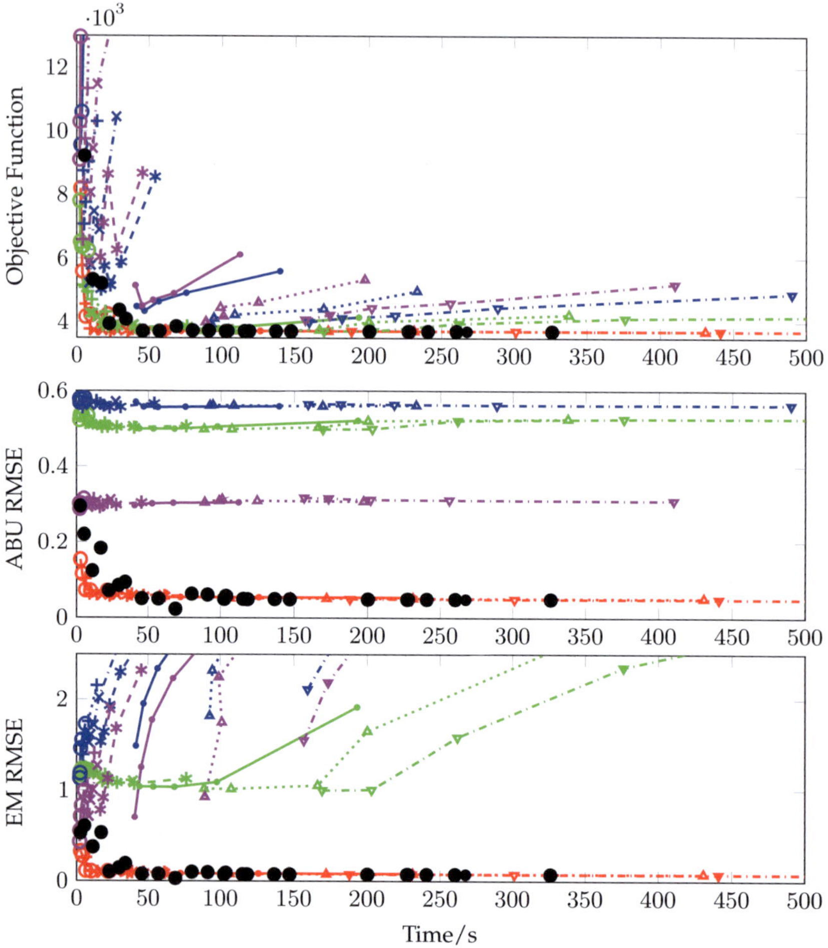

Figure 5.32 Jasper image, 30 dB SNR, PYRU with MiniDisCo: error values over time. Each connected line of data points is composed from data points having the same value of the prefactor V, but different bases B. The legend is the same as in Fig. 5.27.

abundance matrix with random values. This could be more advantageous for the method than reasonably estimated abundance initializations on the ground level that are provided by the unmixing of the first pyramid level already including MiniDisCo.

From this section, the conclusion can be drawn that not every NMF-based unmixing method is equally well suited for the application of PYRU. While the calculation of MVC can be greatly accelerated, PYRU with MiniDisCo fails in providing high-quality unmixing results in faster time.

5.6 3DDCT

After having analyzed the newly developed unmixing methods described in Sec. 4 that are based on local spatial adjacency, it is interesting how 3DDCT-NIMF (3D discrete cosine transform nonnegative inverse matrix factorization) performs. By contrast to UNCLI (unmixing of clustered image) and PYRU (pyramid unmixing), it incorporates the global adjacency of all 3D image elements. All results shown in this section are obtained with the 3DDCT-NIMF procedure described in Sections 4.4.3–4.4.6. As a reminder, the used objective function is (4.41)

$$Q_{\mathrm{3DDCT}} = \sum_{j=1}^{J} \|\mathbf{N}_j \mathbf{Y}_j - \mathbf{A}\mathbf{W}_{XY,j}\|_{\mathrm{F}}^2 - K_j \log(|\det(\mathbf{N}_j)|) \,. \tag{5.8}$$

This means that the non-convex regularizer $-\log(|\det(\mathbf{N}_j)|)$ is used, having the consequence that the outcome of 3DDCT-NIMF depends on the initialization. For this reason, different endmember initializations will be applied, each consisting of R randomly selected pixel spectra. The regularization parameters K_j were given by (4.42):

$$K_j = K \cdot \left(P \frac{\sum_{r=1}^{R} \sum_{i=1}^{I} |\mathbf{Y}_{j,ri}|}{\mathrm{nnz}(\mathbf{Y}_j)} \right)^E \,. \tag{5.9}$$

As these parameters depend on the block index j, the parameters K and E are used as a general substitute. They are the same for each block, which reduces the number of parameters from J to 2. Incorporating each block's given image data, K and E are used to adapt the regularization parameter K_j to each block. For all experiments presented in the following, the constant P is chosen to 10^6.

The regularization parameters K_j have a significant impact on the estimated matrices $\mathbf{A}$ and $\mathbf{N}_j$. The larger K_j, the more emphasis is put on a large determinant, which in turn leads to large absolute elements of the matrices $\mathbf{N}_j$. Afterwards, all these matrices are inverted and the inverse DCT is calculated to get the estimated endmembers $\mathbf{M}$. As the $\mathbf{N}_j$ have very large elements, $\mathbf{M}$ will have quite small elements. In order to minimize the Frobenius norm in (5.8), the abundances will also take high values of similar magnitude as the $\mathbf{N}_j$.

Due to the fact that the image mean was subtracted before the application of 3DDCT-NIMF, the estimated endmembers will be located around 0. In case they are not fully mean-free, the remaining mean of the estimated matrix $\mathbf{M}$ will be subtracted, to have zero mean for both the estimated endmembers and the given image matrix $\mathbf{Y}$ before applying the scaling procedure from Sec. 4.4.5. The result of the scaling is that the endmembers are scaled to the same level as the image pixel spectra. As this procedure provides quite feasible results in the conducted experiments, it will be used in the following. However, it is subject to future work to investigate if it is reasonable to subtract the combined mean of all endmember spectra from the estimated $\mathbf{M}$ or some other value or nothing: an image having mean value 0 (whole image matrix; all pixels) does not require the endmember matrix to also have the mean 0. The image mean value results from the endmembers weighted with the abundances at each pixel, and therefore does not only depend on the endmember spectra and their combined mean.

Figure 5.33 depicts the influence of K on the estimated endmembers. The graphics show the endmembers estimated in the first decomposition step using the full cube, i.e., all J blocks in the top plot of Fig. 4.15, before only the highlighted region in the bottom plot of the same figure is used for the final estimation. This explains why there is some noise present in the estimated endmembers. This noise will be removed when considering only the high-SNR region in the last step. Note that K mainly influences the magnitude of the estimated spectra, but not primarily their shape, whereas the parameter E in conjunction with the initialization changes the endmember shape.

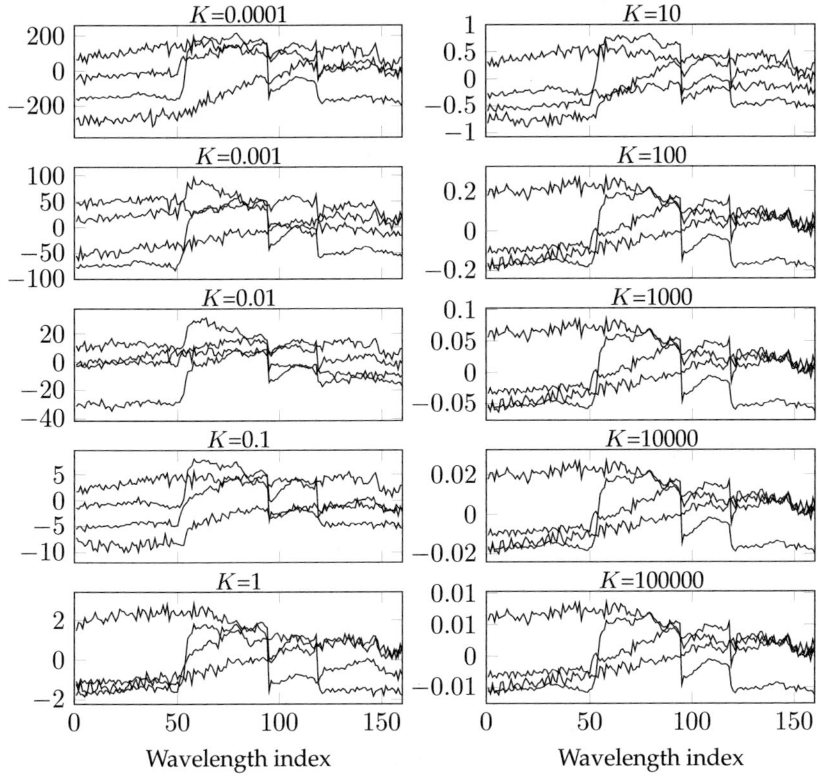

Figure 5.33 Urban Detail4 at 30 dB SNR, $\rho = 0.2$: influence of K on the estimated endmember spectra, shown here before the scaling. The value of K mainly influences the spectra magnitude; the larger K, the smaller the magnitude. Particularly for $K = 100, \ldots, 100000$, the shape of the spectra is nearly the same and does not change with K.

5.6.1 3DDCT-NIMF Unmixing Results of the Remotely Sensed Images

The 3DDCT-NIMF unmixing results of the images described in Sec. 5.1.1 will be illustrated in detail in the following. For all images, 28 different random endmember initializations will be investigated, meaning that for

Table 5.10 Used 3DDCT-NIMF parameter values for each image.

Image name	K	E	T
Urban Detail4	10000	0.1, 0.2,..., 1.2	3000
Urban Detail6	10000	0.1, 0.2,..., 1.2	3000
Jasper	1000	0.1, 0.2,..., 1.2	3000
White Powders	30000	0.05, 0.1, 0.2,..., 1.0	3000
Spatial Mixing 10	30000	0.2, 0,4..., 1.0	3000
Spatial Mixing 20	30000	0.2, 0,4..., 1.0	3000
Spatial Mixing 30	30000	0.2, 0,4..., 1.0	3000

each initialization, R randomly selected pixel spectra compose the initial endmember matrix $\mathbf{M}$. Table 5.10 shows the used parameter values for each image. In addition, the ratio of retained coefficients was altered: $\rho \in \{0.1, 0.2, 0.5\}$. The value $\rho = 0.1$ means that only the largest 10 % of all coefficients are kept and the others are set to zero.

Table 5.11 depicts the 3DDCT-NIMF unmixing results of the Urban Detail4 image. Due to the fact that the ground truth is normally not known, the only way to choose the supposedly best unmixing result is the image reconstruction error, IMAGE RMSE. Apart from the noiseless image, the results show that this error measure provides a feasible criterion for selecting the optimal result; the best ABU RMSE and the best EM RMSE do not differ significantly from the ones of the run that provides the lowest IMAGE RMSE. For the presented analyses, the ground truth is known. For this reason, the column "RMSE type" indicates which error measure (ABU RMSE, EM RMSE, IMAGE RMSE) is the one for which the respective row shows the lowest error values. In the first row, for instance, the result having the lowest IMAGE RMSE is given, along with its ABU RMSE and EM RMSE values. For the image with 30 dB SNR and $\rho = 0.2, 0.5$, the presented run provides the best error values for all the errors at the same time. The column "Init. #" shows which initialization was the one providing the lowest error, so it can be compared if the best ABU RMSE and the best EM RMSE come from the same initialization or not. This makes it possible to distinguish if the result only provides a good abundance estimation and a bad endmember estimation and vice versa, or

Table 5.11 Urban Detail4 image: Error values depending on initialization. The best values of each noise level are highlighted.

	ρ	RMSE type	Init. #	K	E	B	IMAGE	ABU	EM
No noise	0.1	IMAGE	25	10000	0.8	4	0.096	0.214	0.352
		ABU	24	10000	0.9	100	0.170	0.135	0.198
		EM	21	10000	0.8	10	0.146	0.145	0.158
	0.2	IMAGE	27	10000	0.9	6	0.103	0.220	0.309
		ABU	6	10000	0.8	10	0.138	0.151	0.191
		EM	2	10000	0.6	6	0.144	0.154	0.169
	0.5	IMAGE	25	10000	0.8	4	0.117	0.179	0.279
		ABU	6	10000	0.8	10	0.144	0.149	0.165
		EM	6	10000	0.8	10	0.144	0.149	0.165
SNR 20 dB	0.1	IMAGE	24	10000	0.1	30	0.142	0.112	0.141
		ABU	21	10000	0.9	30	0.150	0.099	0.150
		EM	11	10000	0.1	30	0.143	0.112	0.140
	0.2	IMAGE	18	10000	0.1	30	0.142	0.110	0.138
		ABU	2	10000	0.9	20	0.152	0.100	0.148
		EM	19	10000	0.2	30	0.142	0.109	0.130
	0.5	IMAGE	17	10000	0.1	30	0.138	0.090	0.107
		ABU	20	10000	0.7	30	0.163	0.087	0.158
		EM	26	10000	0.1	20	0.138	0.092	0.106
SNR 30 dB	0.1	IMAGE	22	10000	0.9	20	0.086	0.121	0.169
		ABU	26	10000	0.9	100	0.093	0.093	0.126
		EM	26	10000	0.7	50	0.094	0.099	0.122
	0.2	IMAGE	14	10000	0.9	50	0.077	0.077	0.103
		ABU	14	10000	0.9	50	0.077	0.077	0.103
		EM	14	10000	0.9	50	0.077	0.077	0.103
	0.5	IMAGE	21	10000	0.9	30	0.078	0.068	0.086
		ABU	21	10000	0.9	30	0.078	0.068	0.086
		EM	21	10000	0.9	30	0.078	0.068	0.086

if both are of high quality at the same time. The columns K and E show the parameter values of the run that resulted in the best respective error measure.

In the following, a brief discussion of the results given in Tab. 5.11 is conducted. The larger the ratio of retained coefficients ρ, the more coefficients are retained. Its influence is of minor importance; for higher ρ, the error values improve only little. Interestingly, the noisy image with 30 dB SNR provides the lowest error values and not the noiseless one. The 20 dB SNR image leads to worse unmixing results than the 30 dB one.

Figure 5.34 illustrates the best estimated abundance maps of each noise level, while Fig. 5.35 shows the best endmember estimations. A detailed view of some endmembers is provided in Fig. 5.37. The accuracy of both abundance and endmember estimation is quite impressive when considering that the data have been manipulated in a rather rude way: the full image was transformed by 3D DCT; furthermore, the endmembers have been split into blocks and estimated in the inverted domain.

The influence of the parameter B was explained in Sec. 4.4.4. It adapts the width of the column describing which 3D DCT coefficients are used for calculating the matrices $\mathbf{N}_j$. In theory, the smaller B, the less noise corrupted the estimated endmembers should be, as mostly only high-SNR coefficients are considered. The final value of B is selected by analyzing the IMAGE RMSE: for each value of B and the previously calculated abundances, the error between the full image and the reconstruction by the calculated endmembers and abundances is determined. The algorithm decides for the B that provides the lowest reconstruction error. Alternatively, the estimated endmembers of each B can be analyzed manually and a different endmember result can be selected as the final one. Figure 5.36 exemplarily shows the influence of B in case of the optimal endmember estimation resulting from the Urban Detail4 image with 30 dB SNR.

By analogy with Tab. 5.11, Tab. 5.12 shows the error values of the Urban Detail6 image. The unmixing of this image is much more prone to noise than the unmixing of the Urban Detail4 image. The reason is that the number of endmembers $R = 6$ is higher, which means that the horizontal tile in the bottom graphic of Fig. 4.15 is not four layers, but six layers thick. Therefore, more low-SNR coefficients are taken into account which tamper the abundance estimation. For this image, the influence of ρ is

Table 5.12 Urban Detail6 image: Error values depending on initialization. The best values of each noise level are highlighted.

	ρ	RMSE type	Init. #	K	E	B	IMAGE	ABU	EM
No noise	0.1	IMAGE	24	10000	0.9	100	0.117	0.120	0.249
		ABU	24	10000	0.9	100	0.117	0.120	0.249
		EM	26	10000	0.9	100	0.125	0.139	0.231
	0.2	IMAGE	13	10000	0.9	15	0.096	0.141	0.263
		ABU	1	10000	0.9	100	0.127	0.114	0.251
		EM	27	10000	0.9	100	0.110	0.151	0.202
	0.5	IMAGE	8	10000	0.8	10	0.101	0.156	0.362
		ABU	3	10000	0.8	30	0.143	0.132	0.300
		EM	21	10000	0.8	30	0.135	0.162	0.216
SNR 20 dB	0.1	IMAGE	3	10000	0.1	50	0.162	0.151	0.234
		ABU	8	10000	0.8	10	0.173	0.141	0.221
		EM	22	10000	0.8	10	0.172	0.143	0.220
	0.2	IMAGE	13	10000	0.1	30	0.165	0.149	0.246
		ABU	23	10000	0.9	30	0.172	0.138	0.241
		EM	3	10000	0.1	50	0.169	0.140	0.196
	0.5	IMAGE	19	10000	0.1	30	0.163	0.143	0.234
		ABU	15	10000	0.9	30	0.180	0.133	0.221
		EM	3	10000	0.2	30	0.171	0.160	0.217
SNR 30 dB	0.1	IMAGE	28	10000	0.9	100	0.119	0.119	0.210
		ABU	5	10000	0.7	100	0.125	0.116	0.168
		EM	25	10000	0.7	100	0.128	0.117	0.163
	0.2	IMAGE	10	10000	0.3	50	0.110	0.123	0.200
		ABU	25	10000	0.5	50	0.114	0.098	0.161
		EM	14	10000	0.1	50	0.120	0.118	0.129
	0.5	IMAGE	5	10000	0.3	50	0.109	0.119	0.138
		ABU	25	10000	0.8	10	0.119	0.089	0.177
		EM	14	10000	0.1	50	0.111	0.121	0.125

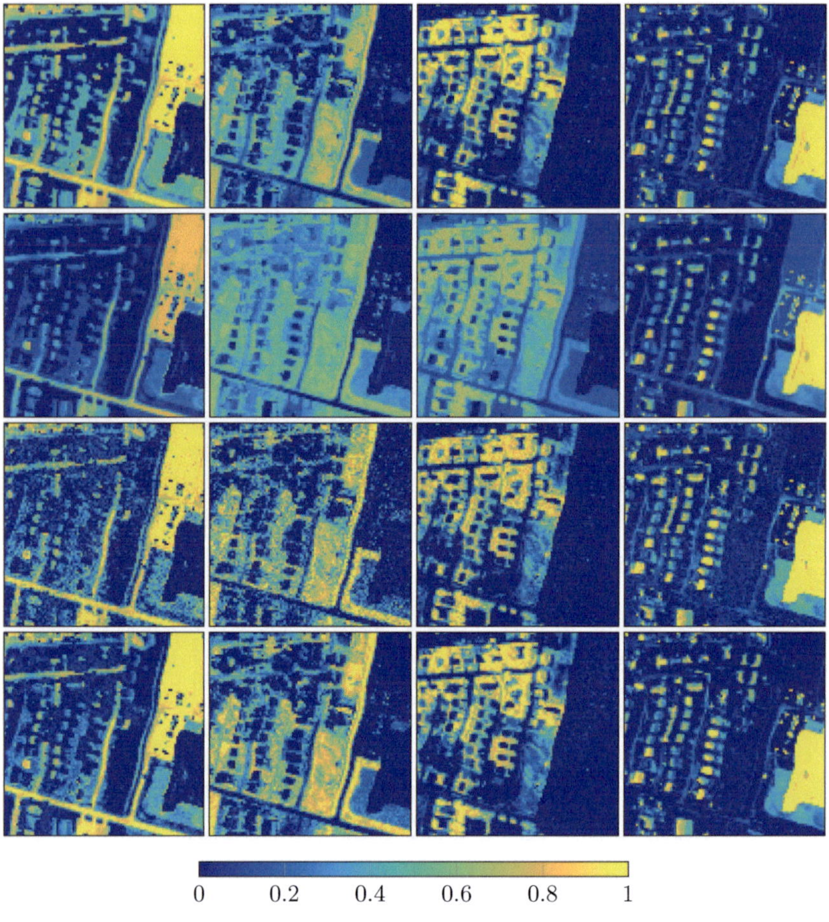

Figure 5.34 Urban Detail4 image: resulting best abundance maps. Top row: ground truth, second row: no noise, third row: 20 dB SNR, last row: 30 dB SNR. For all noise levels, the spatial structure is preserved really well, however, in the noiseless case, three abundance maps are estimated with the wrong scale. For 20 dB SNR, there is some noise in the estimation visible.

even lower than for the Urban Detail4 image. Figures 5.38 and 5.39 depict the abundance and endmember estimation in the 30 dB SNR scenario.

Table 5.13 Jasper image: Error values depending on initialization. The best values of each noise level are highlighted.

	ρ	RMSE type	Init. #	K	E	B	IMAGE	ABU	EM
No noise	0.1	IMAGE	23	1000	0.8	10	0.209	0.257	0.814
		ABU	4	1000	1.2	100	0.224	0.202	1.308
		EM	13	1000	0.7	8	0.240	0.220	0.532
	0.2	IMAGE	4	1000	1.2	100	0.196	0.209	1.138
		ABU	12	1000	1.2	100	0.237	0.197	0.665
		EM	1	1000	0.7	4	0.286	0.238	0.426
	0.5	IMAGE	4	1000	1.2	50	0.216	0.212	1.135
		ABU	28	1000	1.2	100	0.224	0.204	0.579
		EM	1	1000	0.8	4	0.288	0.256	0.447
SNR 20 dB	0.1	IMAGE	10	1000	1.1	50	0.197	0.182	0.352
		ABU	5	1000	0.8	50	0.251	0.156	0.267
		EM	15	1000	0.4	30	0.235	0.157	0.206
	0.2	IMAGE	10	1000	1.1	50	0.202	0.190	0.328
		ABU	6	1000	0.6	30	0.238	0.145	0.266
		EM	5	1000	0.3	30	0.220	0.147	0.221
	0.5	IMAGE	8	1000	0.1	20	0.207	0.192	0.367
		ABU	25	1000	1.2	50	0.260	0.142	0.382
		EM	28	1000	0.2	10	0.214	0.148	0.229
SNR 30 dB	0.1	IMAGE	6	1000	0.9	100	0.140	0.128	0.214
		ABU	2	1000	0.9	100	0.159	0.123	0.324
		EM	6	1000	0.6	50	0.145	0.128	0.209
	0.2	IMAGE	22	1000	1.2	100	0.134	0.127	0.260
		ABU	2	1000	0.7	100	0.155	0.108	0.264
		EM	25	1000	0.3	50	0.147	0.125	0.230
	0.5	IMAGE	27	1000	0.2	20	0.141	0.219	0.578
		ABU	2	1000	0.6	50	0.157	0.110	0.221
		EM	2	1000	0.6	50	0.157	0.110	0.221

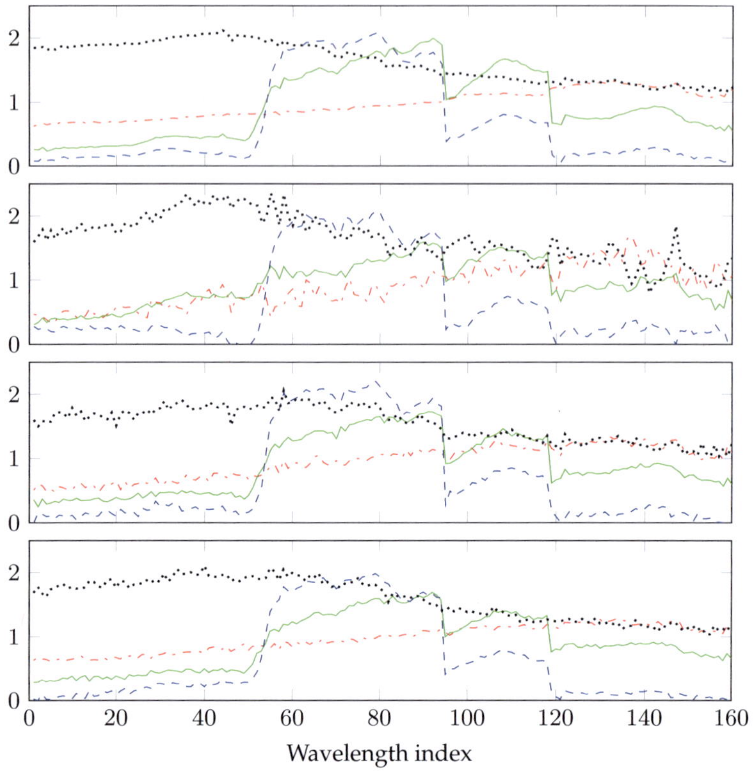

Figure 5.35 Urban Detail4 image: best estimated endmember spectra. Top row: ground truth, second row: no noise, third row: 20 dB SNR, last row: 30 dB SNR. Note the detailedness of the estimation; see also Fig. 5.37 for a more detailed display.

The unmixing results of the Jasper image are shown in Tab. 5.13. In comparison with the Urban Detail4 image, it has the same number of endmembers R and the same spatial size, but a higher number of wavelengths L. Its decomposition results in much higher error values than the Urban Detail4 image. While the main structure of both endmembers and abundances is preserved, large noise levels are present. The reason for this is unknown; changing the exponent values, the number of iterations and

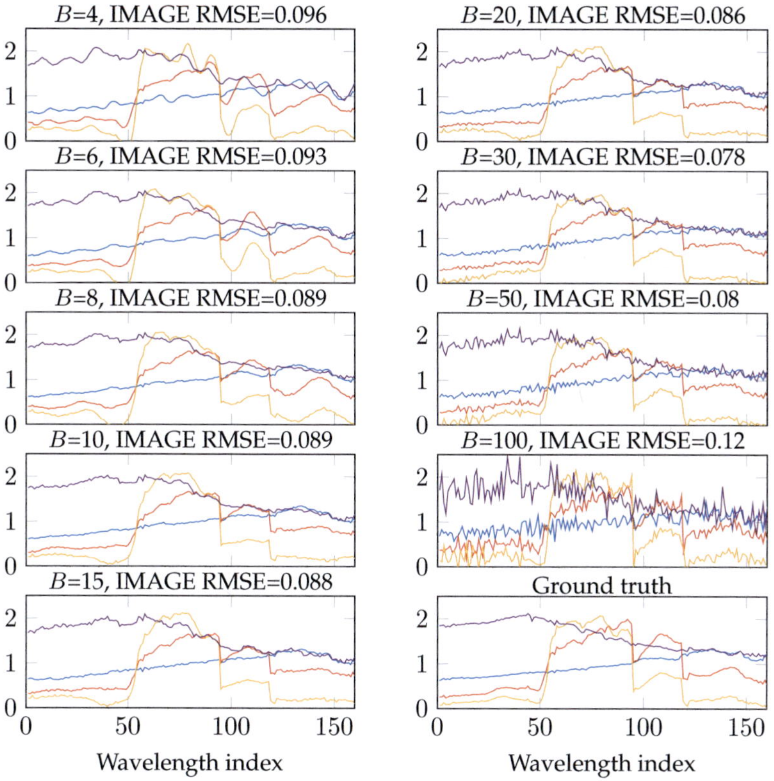

Figure 5.36 Urban Detail4 at 30 dB SNR, $\rho = 0.5$: dependence of the estimated spectra on B. The larger B, the more details appear, but the noise variance also increases. The optimal B leading to the lowest IMAGE RMSE is 30.

running more initializations did not lead to an improvement.

5.6.2 3DDCT-NIMF Unmixing Results of the Lab Images

In the previous section, only remotely sensed images have been analyzed. They have been processed by the image providers such that they can be considered as noise-free, and before the analysis, AWGN with known SNR has been added. As a consequence, the noise present in each 3D image

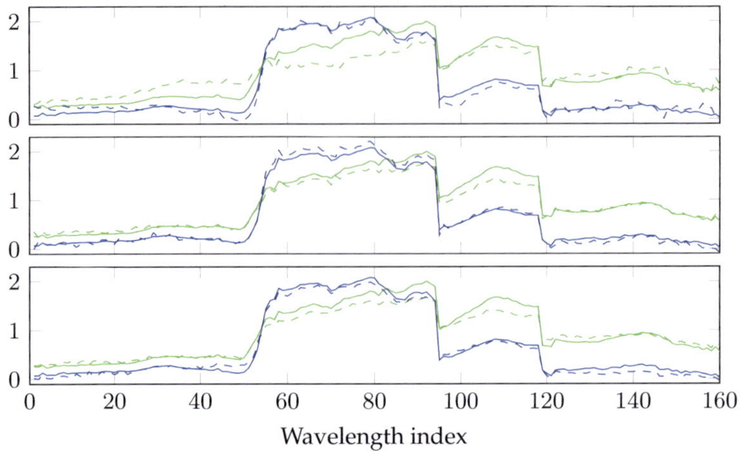

Figure 5.37 Urban Detail4 image: details of two of the estimated spectra depicted in Fig. 5.35. Within each plot, the solid line denotes the ground truth and the dash-dotted line the estimation. Top row: no noise, second row: 20 dB SNR, last row: 30 dB SNR. Especially for the noisy images, the level of detailedness is remarkable. As indicated in Tab. 5.11, the optimal B is smaller than the image's spatial dimensions, so low-SNR DCT coefficients have been disregarded. See also the vertical column in the bottom plot of Fig. 4.15.

element is fully characterized. By contrast, the images White Powders as well as Spatial Mixing 10, Spatial Mixing 20 and Spatial Mixing 30 have been acquired directly without further processing. No analysis of the noise characteristics of the used camera was performed, as this would be out of the scope of this thesis. On the contrary, apart from performing the white balance described in Sec. 5.1.1 and correcting defective pixels, the images are used as acquired. The goal is to find out to which extent 3DDCT-NIMF can deal with non-Gaussian noise with variable variance in each image element. This is in contrast to the noise considerations discussed in Sec. 4.4.4, where it was assumed that the noise is Gaussian and its variance is constant throughout all image elements.

The White Powders image is hard to unmix, especially for the 3DDCT-NIMF method. It has a comparatively large number of endmembers, all ground truth endmembers are quite close together, and the image is of

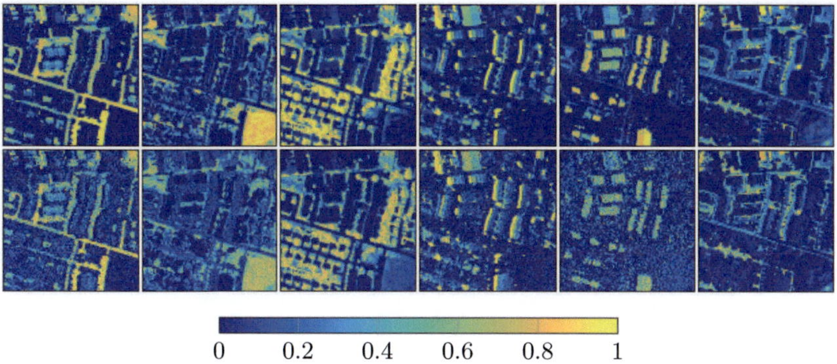

Figure 5.38 Urban Detail6 image, SNR 30 dB: best estimated abundances. Top row: ground truth, second row: estimation.

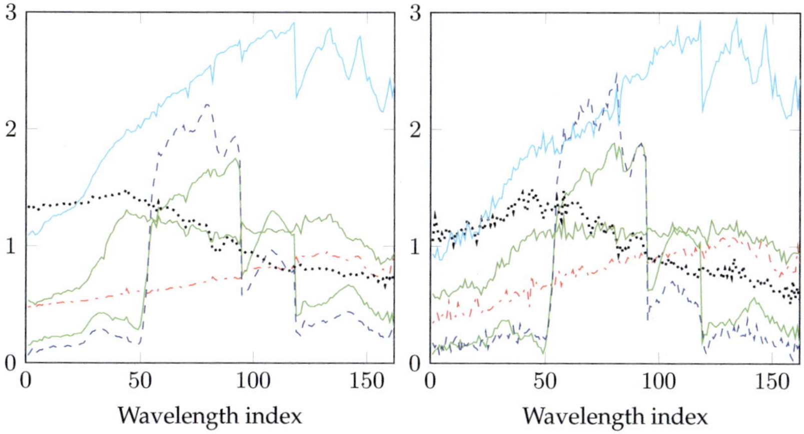

Figure 5.39 Urban Detail6 image, SNR 30 dB: best estimated endmember spectra. Left: ground truth, right: estimation.

large spatial size. Apart from this, the noise characteristics of the image are unknown. Nevertheless, the 3DDCT-NIMF unmixing results are of remarkable quality, see Figures 5.40, 5.41 and 5.42. Especially the unmixing errors that are depicted in Tab. 5.14 can compete with MVC (see Sec. 5.5.1 for PYRU and Tab. 5.8 for UNCLI) and MiniDisCo (see Sec. 5.5.2 for PYRU

Table 5.14 White Powders image: Error values depending on initialization. The best values are highlighted.

ρ	RMSE type	Init. #	K	E	B	IMAGE	ABU	EM
	IMAGE	22	30000	0.05	50	0.021	0.132	0.905
0.1	ABU	11	30000	0.2	320	0.032	0.130	1.158
	EM	25	30000	0.3	320	0.050	0.146	0.232
	IMAGE	22	30000	0.05	320	0.021	0.141	0.784
0.2	ABU	9	30000	0.1	30	0.069	0.130	0.338
	EM	25	30000	0.05	50	0.049	0.144	0.240
	IMAGE	21	30000	0.05	320	0.021	0.162	0.864
0.5	ABU	27	30000	0.05	50	0.073	0.130	1.364
	EM	25	30000	0.3	50	0.052	0.152	0.203

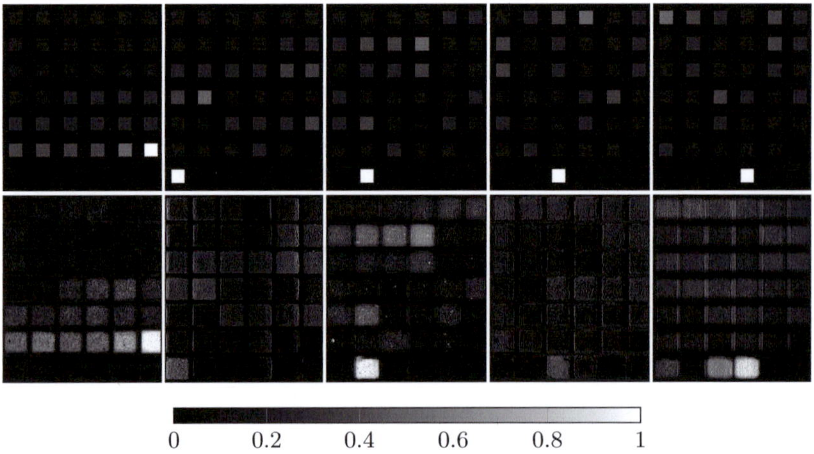

Figure 5.40 White Powders image: best resulting abundance maps. Top row: ground truth, second row: estimation. The abundance of the box frame which would be the sixth substance is not shown. Remember that the error is only calculated within the areas for which at least one abundance is > 0. Although the visual impression of the estimation could exhibit a larger degree of correspondence, the ABU RMSE values are better than the one obtained with other methods, see Tab. 5.18.

Table 5.15 Spatial Mixing images: Error values depending on initialization. The best values of each image are highlighted.

	ρ	RMSE type	Init. #	K	E	B	IMAGE	ABU	EM
Spatial Mixing 10	0.1	IMAGE	4	30000	0.6	200	0.350	0.197	1.024
		ABU	1	30000	0.8	200	0.416	0.175	0.691
		EM	2	30000	0.4	200	0.391	0.186	0.644
	0.2	IMAGE	20	30000	0.2	200	0.346	0.198	0.998
		ABU	9	30000	0.8	50	0.398	0.171	0.780
		EM	17	30000	0.2	200	0.408	0.193	0.612
	0.5	IMAGE	21	30000	0.4	200	0.338	0.229	1.684
		ABU	8	30000	0.8	200	0.416	0.155	0.711
		EM	15	30000	0.2	200	0.392	0.176	0.568
Spatial Mixing 20	0.1	IMAGE	26	30000	0.6	147	0.306	0.194	0.744
		ABU	18	30000	0.8	50	0.362	0.172	0.645
		EM	23	30000	0.4	147	0.336	0.181	0.599
	0.2	IMAGE	22	30000	0.4	147	0.290	0.200	0.899
		ABU	23	30000	0.8	147	0.380	0.165	0.619
		EM	4	30000	0.6	147	0.382	0.182	0.570
	0.5	IMAGE	26	30000	0.4	147	0.284	0.213	1.075
		ABU	1	30000	0.8	147	0.358	0.171	0.620
		EM	10	30000	0.4	147	0.340	0.192	0.601
Spatial Mixing 30	0.1	IMAGE	3	30000	0.4	98	0.252	0.206	0.775
		ABU	28	30000	0.8	98	0.308	0.168	0.666
		EM	28	30000	0.4	98	0.286	0.171	0.632
	0.2	IMAGE	28	30000	0.4	98	0.248	0.197	0.778
		ABU	28	30000	0.8	98	0.305	0.170	0.675
		EM	14	30000	0.8	98	0.301	0.178	0.605
	0.5	IMAGE	28	30000	0.6	98	0.234	0.206	0.861
		ABU	27	30000	0.8	98	0.319	0.174	0.782
		EM	27	30000	0.4	98	0.272	0.182	0.633

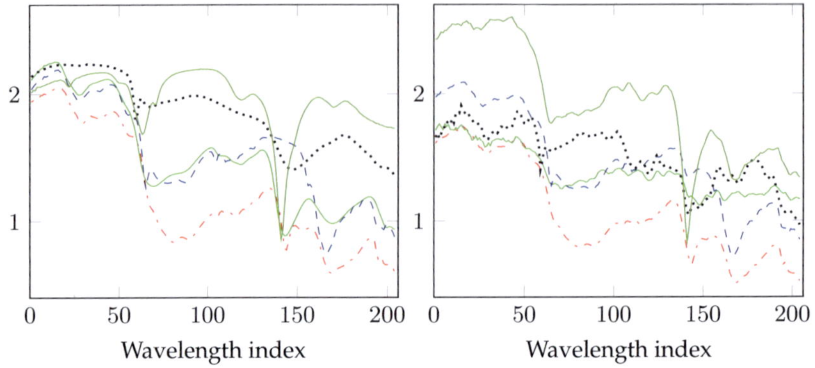

Figure 5.41 White Powders image: best estimated endmember spectra. Left: ground truth, right: estimation. Frame spectrum not shown. The overall shape for most spectra is estimated quite well; major problem is the correct scaling.

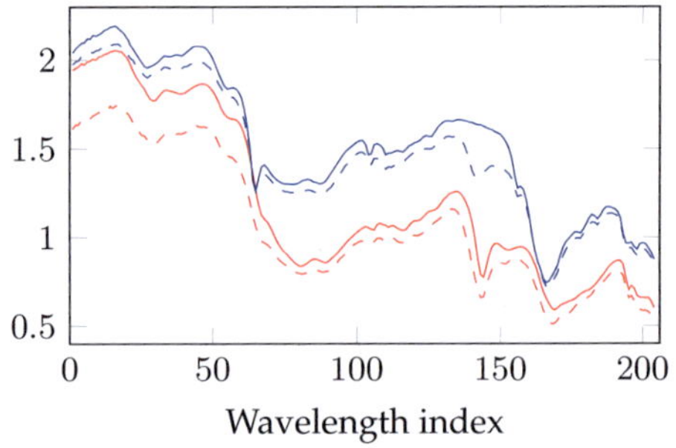

Figure 5.42 White Powders image: details of two of the estimated spectra depicted in Fig. 5.41. Solid line denotes ground truth, dash-dotted line estimation.

and Tab. 5.9 for UNCLI). In comparison with the other images, the ABU RMSE is even more stable, i.e., for the run providing the best IMAGE RMSE/EM RMSE, the ABU RMSE is close to its best value. The opposite holds for the EM RMSE.

The unmixing errors of the Spatial Mixing images are given in Tab. 5.15. All images provide relatively similar error values, also quite independent of the ratio ρ. The error values are in the range of the other methods, see Tables 5.8 and 5.9.

Table 5.18 in the following section provides a final comparison between the newly developed 3DDCT-NIMF and 3DSVD-NIMF methods as well as MVC and MiniDisCo from the literature.

5.7 3DSVD and Final Comparison

The 3D singular value nonnegative inverse matrix factorization (3DSVD-NIMF) method introduced in Sec. 4.5 is applied to the hyperspectral images explained in Sec. 5.1.1. For all investigations performed with 3DSVD-NIMF, the parameter K was chosen to 10000 and the exponents $E = 0.1, 0.2, \ldots, 1.5$ were tried. The number T of iterations is 3000 and 28 initializations were run for each image. By contrast to 3D discrete cosine transform nonnegative inverse matrix factorization (3DDCT-NIMF), a higher ratio ρ of kept coefficients is required to adequately represent the image. Note that the ratio ρ refers to all coefficients of the full cube, but only the first few blocks are used for calculating abundances and endmembers according to the objective function (4.41). The value of ρ is set to 0.9 for all images. The resulting 3DSVD-NIMF error values are shown in Tab. 5.16. By contrast to the analyses presented for 3DDCT-NIMF (Sec. 5.6), only the best values are given and not their parameters. On the one hand, this is for the sake of brevity, on the other hand, this is due to one remarkable property: the unmixing quality hardly depends on the endmember initialization, as long as at least 2 blocks are considered. Table 5.17 shows a comparison between 3DDCT-NIMF and 3DSVD-NIMF. For each method, the same number of blocks was used and the influence of the initialization is illustrated. For the parameter combination that provides the best ABU RMSE, the difference between maximum and minimum ABU RMSE for all 28 initializations is provided.

3DDCT-NIMF and 3DSVD-NIMF are compared with MVC and MiniDisCo in Tab. 5.18. For MVC and MiniDisCo, the results of full image unmixing are given, PYRU and UNCLI were not used. While MVC provides for most remotely sensed images the best unmixing quality, 3DDCT-NIMF and 3DSVD-NIMF provide similar results for the Urban Detail6 image with 30 dB SNR. For all laboratory images, 3DDCT-NIMF provides the better endmember estimation and for some of them also the best abundance estimation. This comparison shows that while 3DDCT-NIMF is

Table 5.16 3DSVD-NIMF results when only the first few blocks are considered. The columns show the best ABU RMSE and the best EM RMSE for each image.

	1 block		2 blocks		3 blocks		5 blocks	
Image	ABU	EM	ABU	EM	ABU	EM	ABU	EM
Jasper								
20 dB	0.120	0.309	**0.083**	**0.134**	0.084	0.140	0.086	0.148
30 dB	0.108	0.262	**0.065**	**0.109**	0.067	0.118	0.070	0.127
Urban Detail4								
20 dB	0.112	0.170	**0.073**	**0.053**	0.076	0.060	0.077	0.069
30 dB	0.079	0.112	**0.066**	**0.057**	0.069	0.062	0.071	0.066
Urban Detail6								
20 dB	0.163	0.364	**0.093**	**0.156**	0.093	0.165	0.093	0.170
30 dB	0.129	0.261	**0.056**	**0.133**	0.058	0.146	0.060	0.155
White Powders	–	–	0.126	0.226	–	–	–	–

Table 5.17 Dependence of 3DDCT-NIMF and 3DSVD-NIMF on the initialization. For the parameter combination that provides the lowest ABU RMSE, the difference between maximum and minimum ABU RMSE among all initializations is given. Note that for more than one block, all 3DSVD-NIMF initializations lead to almost the same ABU RMSE, as the difference between maximum and minimum is very small, see the last column.

	1 block		2 blocks	
Image	3DDCT	3DSVD	3DDCT	3DSVD
Jasper 20 dB	0.1598	0.0773	0.1464	0.0006
Jasper 30 dB	0.0834	0.1616	0.1400	0.0004
Urban Detail4 20 dB	0.0673	0.0620	0.2077	0.0012
Urban Detail4 30 dB	0.0898	0.1228	0.0833	0.0010
Urban Detail6 20 dB	0.0285	0.0260	0.0431	0.0003
Urban Detail6 30 dB	0.0409	0.0550	0.0443	0.0002

Table 5.18 Comparison between MVC, MiniDisCo (full image unmixing, no UNCLI or PYRU), 3DDCT-NIMF and 3DSVD-NIMF unmixing results. The given numbers are the RMSE values according to the column heading. The MVC values are taken from Tab. 5.8 and the MiniDisCo ones from Tab. 5.9. For 3DDCT-NIMF, the results are the best ones of the Tables 5.14, 5.11, 5.12, 5.14 and 5.15; the shown EM RMSE is the one that belongs to the optimal ABU RMSE and not the optimal one. The same holds for 3DSVD-NIMF, for which the results are taken from Tab. 5.16.

	MVC		MiniDisCo		3DDCT		3DSVD	
Image	ABU	EM	ABU	EM	ABU	EM	ABU	EM
Jasper								
No noise	**0.007**	**0.011**	0.046	0.069	0.197	0.665	–	–
20 dB	**0.021**	**0.020**	0.055	0.085	0.142	0.382	0.083	0.134
30 dB	**0.009**	**0.012**	0.049	0.074	0.108	0.264	0.065	0.110
Urban Detail4								
No noise	**0.011**	**0.013**	0.062	0.066	0.135	0.198	–	–
20 dB	**0.022**	**0.015**	0.066	0.066	0.087	0.158	0.073	0.067
30 dB	**0.013**	**0.013**	0.062	0.066	0.068	0.086	0.066	0.058
Urban Detail6								
No noise	**0.039**	**0.057**	0.095	0.171	0.114	0.251	–	–
20 dB	**0.083**	0.192	0.103	**0.144**	0.133	0.221	0.093	0.158
30 dB	**0.053**	0.152	0.102	0.184	0.089	0.177	0.056	**0.144**
White Powders	0.169	0.485	0.207	0.809	0.130	**0.338**	**0.126**	0.414
Spatial Mixing 10	0.173	1.023	0.193	1.187	**0.155**	**0.711**	–	–
Spatial Mixing 20	0.195	1.006	**0.158**	1.197	0.165	**0.619**	–	–
Spatial Mixing 30	0.162	1.128	**0.141**	1.082	0.168	**0.666**	–	–

not always able to provide state-of-the-art results, it comes close to them; 3DSVD-NIMF even closer. 3DSVD-NIMF is characterized by extraordinary robustness to initialization. In general, it is remarkable that applying the 3D DCT or 3D SVD and using inverted endmember matrices provide such good results at all.

6 Conclusion

6.1 Summary

Several new unmixing methods and a new multidimensional image denoising algorithm have been presented in this thesis. The unmixing of clustered image, UNCLI, is able to calculate less accurate results than full image unmixing in shorter time or more accurate results in longer time, in a considerable number of cases even better results in shorter time. The influence of the clustering method on the results has been investigated. Among these methods, RHSEG (recursive hierarchical segmentation) proved to be the most suited one. Especially applying the last step for abundance improvement leads to fast, accurate unmixing.

By using image pyramids applied to hyperspectral images, PYRU (pyramid unmixing) does not aim at better unmixing results, but simply faster calculation. It can be combined with most unmixing methods; here, the usage of unmixing methods based on NMF (nonnegative matrix factorization) has been shown. As long as the chosen unmixing method is suitable and the image consists of linearly mixed spectra with little endmember variability, PYRU can speed up the unmixing by several times. However, not every method can be combined with PYRU. An example for a not feasible method is MiniDisCo (minimum dispersion constrained NMF).

The probably most interesting method is 3DDCT-NIMF, 3D discrete cosine transform nonnegative inverse matrix factorization. It applies a 3D transform, the 3D DCT, to a hyperspectral image. Although this seems to be a natural choice, most existing methods focus on the pixel spectra exploiting their geometrical properties in the L-dimensional space, but neglecting the spatio-spectral adjacency. By contrast, 3DDCT-NIMF incorporates full spatial and spectral adjacency. The use of inverse matrices leads to a convex data fidelity term, this is also by contrast to the vast majority of existing unmixing methods. A very beneficial effect is how the orthogonal transform distributes the signal energy in the transform domain. It is reasonable to concentrate it only in the part of the domain that is required for calculating endmembers and abundances. Although this is not fully exploited yet, 3DDCT-NIMF is able to provide results that are close to the ones of existing unmixing methods.

Applying the singular value decomposition (SVD) analogously to the DCT leads to 3D singular value nonnegative matrix factorization (3DSVD-NIMF). It provides slightly better results than 3DDCT-NIMF. Its most interesting fact that its outcome is almost not influenced by random initialization.

The unmixing methods have been applied to simulated sorting images imitating a belt sorting scenario. Although the linear mixing model holds in this case, endmember variability has a significant impact and should be considered. Further investigations should focus on implementing spectral unmixing in real sorters. Difficulties are the data processing hardware and software, as hyperspectral images have high demand on both disk space and processing time.

Although it is a relatively simple denoising algorithm, MIDWAI (multidimensional image denoising by weighted addition of images) is quite powerful in denoising multidimensional images. This has been shown with three-dimensional hyperspectral reflectance images as well as four-dimensional hyperspectral fluorescence images. The results are in the range of state-of-the-art algorithms, while far less calculation time is required for good denoising quality. MIDWAI versions for both Poisson and Gaussian noise have been presented.

6.2 Outlook

From the methods and results presented in this thesis, many further research directions can be proposed. It is definitely interesting to apply 3DDCT-NIMF to other types of data. The objective function is the same in many different applications such as audio signal processing for source separation and text analysis and many more. Although these most likely do not have an inherent 3D structure as hyperspectral images, the data contained in the 2D data matrix $\mathbf{Y}$ can be arranged to a cube and a 3D transform can be applied for obtaining favorable SNR properties. A 3D consideration is likely to lead to a larger sparsity. The chosen orthogonal transform (3D DCT) is also an issue that should be studied in detail: the useful SNR properties of the presented method are not limited to the DCT; any orthogonal method is suitable. Possible candidates are, e.g., the wavelet transform and maximum noise fraction (MNF) [52]. These methods can be applied to all three dimensions, not only to the spectral one; PCA and MNF commonly only are used for spectral dimensionality reduction. It is also possible to construct or learn orthogonal transforms that are especially designed for the image under investigation. Although the theoretical considerations are performed with AWGN with constant variance in each image element, it is likely and has been shown with the White Powders dataset that the considerations to a certain extent also hold for other types of noise with varying variance in each image element. One natural next step could be the incorporation of the convex regularizer ensuring matrix invertibility that was presented in Sec. 4.4.7 with minimum volume regularizers. This may lead to a fully convex objective function.

It is also a natural continuation of the work done so far to exploit the signal energy properties described in the 3DDCT-NIMF derivation for image or signal denoising in general.

On a more physical than algorithmic level, it is possible to implement the unmixing optically by using suitably designed spectral filters. This possibly circumvents the need for elaborate hardware; especially on the software side, significant reduction in the computational load is possible, as the acquired data are directly the sought-after abundance values. A research project building on the results of this thesis and focusing on developing signal processing algorithms that can be applied optically already has been started. First investigations regarding the determination of the spectral filter curves have already been published [179–182].

Appendix

A 3DDCT-NIMF Gradients with Weighting

Calculating the gradient of the weighted objective function (4.45) requires this relation:

$$\operatorname{tr}\left((\underbrace{\mathbf{A}}_{\in\mathbb{C}^{n\times m}} \odot \underbrace{\mathbf{B}}_{\in\mathbb{C}^{n\times m}}) \underbrace{\mathbf{C}^{\mathrm{T}}}_{\in\mathbb{C}^{m\times n}} \right) = \operatorname{tr}\left((\mathbf{A} \odot \mathbf{C})\mathbf{B}^{\mathrm{T}} \right) . \tag{A.1}$$

Proof:

$$\begin{aligned}
\left((\mathbf{A} \odot \mathbf{B})\mathbf{C}^{\mathrm{T}} \right)_{ik} &= \sum_{j=1}^{m} (\mathbf{A} \odot \mathbf{B})_{ij} \cdot \mathbf{C}^{\mathrm{T}}_{jk} = \sum_{j=1}^{m} a_{ij} \cdot b_{ij} \cdot c_{kj} \\
&\Rightarrow \left((\mathbf{A} \odot \mathbf{B})\mathbf{C}^{\mathrm{T}} \right)_{zz} = \sum_{j=1}^{m} a_{zj} \cdot b_{zj} \cdot c_{zj} \\
&\Rightarrow \operatorname{tr}\left((\mathbf{A} \odot \mathbf{B})\mathbf{C}^{\mathrm{T}} \right) = \sum_{z=1}^{n} \sum_{j=1}^{m} a_{zj} \cdot b_{zj} \cdot c_{zj} ,
\end{aligned} \tag{A.2}$$

$$\begin{aligned}
\left((\mathbf{A} \odot \mathbf{C})\mathbf{B}^{\mathrm{T}} \right)_{ik} &= \sum_{j=1}^{m} (\mathbf{A} \odot \mathbf{C})_{ij} \cdot \mathbf{B}^{\mathrm{T}}_{jk} = \sum_{j=1}^{m} a_{ij} \cdot c_{ij} \cdot b_{kj} \\
&\Rightarrow \left((\mathbf{A} \odot \mathbf{C})\mathbf{B}^{\mathrm{T}} \right)_{zz} = \sum_{j=1}^{m} a_{zj} \cdot b_{zj} \cdot c_{zj} \\
&\Rightarrow \operatorname{tr}\left((\mathbf{A} \odot \mathbf{C})\mathbf{B}^{\mathrm{T}} \right) = \sum_{z=1}^{n} \sum_{j=1}^{m} a_{zj} \cdot b_{zj} \cdot c_{zj} .
\end{aligned} \tag{A.3}$$

One single summand of the objective function (4.45) can be stated differently:

$$
\begin{aligned}
Q &= \|(\mathbf{N}_j\mathbf{Y}_j) \odot \mathbf{V}_j - (\mathbf{A}\mathbf{W}_{xy,j}) \odot \mathbf{V}_j\|_\mathrm{F}^2 \\
&= \mathrm{tr}\Big(\big((\mathbf{N}_j\mathbf{Y}_j) \odot \mathbf{V}_j - (\mathbf{A}\mathbf{W}_{xy,j}) \odot \mathbf{V}_j\big) \\
&\quad \cdot \big((\mathbf{N}_j\mathbf{Y}_j) \odot \mathbf{V}_j - (\mathbf{A}\mathbf{W}_{xy,j}) \odot \mathbf{V}_j\big)^\mathrm{T}\Big) \\
&= \mathrm{tr}\Big(\big((\mathbf{N}_j\mathbf{Y}_j) \odot \mathbf{V}_j\big)\big((\mathbf{N}_j\mathbf{Y}_j) \odot \mathbf{V}_j\big)^\mathrm{T}\Big) \\
&\quad - \mathrm{tr}\Big(\big((\mathbf{N}_j\mathbf{Y}_j) \odot \mathbf{V}_j\big)\big((\mathbf{A}\mathbf{W}_{xy,j}) \odot \mathbf{V}_j\big)^\mathrm{T}\Big) \\
&\quad - \mathrm{tr}\Big(\big((\mathbf{A}\mathbf{W}_{xy,j}) \odot \mathbf{V}_j\big)\big((\mathbf{N}_j\mathbf{Y}_j) \odot \mathbf{V}_j\big)^\mathrm{T}\Big) \\
&\quad + \mathrm{tr}\Big(\big((\mathbf{A}\mathbf{W}_{xy,j}) \odot \mathbf{V}_j\big)\big((\mathbf{A}\mathbf{W}_{xy,j}) \odot \mathbf{V}_j\big)^\mathrm{T}\Big) \\
&= \mathrm{tr}\Big(\big((\mathbf{N}_j\mathbf{Y}_j) \odot \mathbf{V}_j\big)\big((\mathbf{N}_j\mathbf{Y}_j) \odot \mathbf{V}_j\big)^\mathrm{T}\Big) \\
&\quad - 2\,\mathrm{tr}\Big(\big((\mathbf{N}_j\mathbf{Y}_j) \odot \mathbf{V}_j\big)\big((\mathbf{A}\mathbf{W}_{xy,j}) \odot \mathbf{V}_j\big)^\mathrm{T}\Big) \\
&\quad + \mathrm{tr}\Big(\big((\mathbf{A}\mathbf{W}_{xy,j}) \odot \mathbf{V}_j\big)\big((\mathbf{A}\mathbf{W}_{xy,j}) \odot \mathbf{V}_j\big)^\mathrm{T}\Big) .
\end{aligned}
\tag{A.4}
$$

The gradient of one summand from (4.45) with respect to **A** results to:

$$
\begin{aligned}
\frac{\partial Q}{\partial \mathbf{A}} &= \frac{\partial}{\partial \mathbf{A}}\mathrm{tr}\Big(\big((\mathbf{N}_j\mathbf{Y}_j) \odot \mathbf{V}_j\big)\big((\mathbf{N}_j\mathbf{Y}_j) \odot \mathbf{V}_j\big)^\mathrm{T}\Big) \\
&\quad - 2\frac{\partial}{\partial \mathbf{A}}\mathrm{tr}\Big(\big((\mathbf{N}_j\mathbf{Y}_j) \odot \mathbf{V}_j\big)\big((\mathbf{A}\mathbf{W}_{xy,j}) \odot \mathbf{V}_j\big)^\mathrm{T}\Big) \\
&\quad + \frac{\partial}{\partial \mathbf{A}}\mathrm{tr}\Big(\big((\mathbf{A}\mathbf{W}_{xy,j}) \odot \mathbf{V}_j\big)\big((\mathbf{A}\mathbf{W}_{xy,j}) \odot \mathbf{V}_j\big)^\mathrm{T}\Big) \\
&= -2\frac{\partial}{\partial \mathbf{A}}\mathrm{tr}\Big(\big((\mathbf{N}_j\mathbf{Y}_j) \odot \mathbf{V}_j\big)\big((\mathbf{A}\mathbf{W}_{xy,j}) \odot \mathbf{V}_j\big)^\mathrm{T}\Big) \\
&\quad + \frac{\partial}{\partial \mathbf{A}}\mathrm{tr}\Big(\big((\mathbf{A}\mathbf{W}_{xy,j}) \odot \mathbf{V}_j\big)\big((\mathbf{A}\mathbf{W}_{xy,j}) \odot \mathbf{V}_j\big)^\mathrm{T}\Big) .
\end{aligned}
\tag{A.5}
$$

According to [130], the gradient of a scalar function depending on a matrix trace is given by

$$
D_\mathbf{K} f(\mathbf{X}_0) = \lim_{t \to 0} \frac{f(\mathbf{X}_0 + t \cdot \mathbf{K}) - f(\mathbf{X}_0)}{t}
\tag{A.6}
$$

which yields then at the end

$$D_{\mathbf{K}} f(\mathbf{X}_0) = \operatorname{tr}\Big(\mathbf{K}^{\mathrm{T}}\mathbf{U}\Big) \text{ with } \operatorname{tr}\Big(\mathbf{K}^{\mathrm{T}}\mathbf{K}\Big) = 1\,. \tag{A.7}$$

The matrix $\mathbf{U}$ in this case denotes the gradient

$$\mathbf{U} = \frac{\partial f}{\partial \mathbf{X}}\,, \tag{A.8}$$

evaluated at $\mathbf{X}_0$. By calculating the limit in (A.6) and comparing the result with (A.7), the gradient of a trace of a matrix-valued function can be calculated. Applying to (A.5) yields

$$\begin{aligned}
&D_{\mathbf{K}} f(\mathbf{A}_0) \qquad\qquad (A.9)\\
&= \lim_{t\to 0} \frac{\operatorname{tr}\Big(-2\left((\mathbf{N}_j\mathbf{Y}_j)\odot\mathbf{V}_j\right)\left(((\mathbf{A}_0 + t\mathbf{K})\mathbf{W}_{xy,j})\odot\mathbf{V}_j\right)^{\mathrm{T}}\Big)}{t}\\
&\quad + \lim_{t\to 0} \frac{\operatorname{tr}\Big(\left(((\mathbf{A}_0 + t\mathbf{K})\mathbf{W}_{xy,j})\odot\mathbf{V}_j\right)\left(((\mathbf{A}_0 + t\mathbf{K})\mathbf{W}_{xy,j})\odot\mathbf{V}_j\right)^{\mathrm{T}}\Big)}{t}\\
&\quad + 2\lim_{t\to 0} \frac{\operatorname{tr}\Big(\left((\mathbf{N}_j\mathbf{Y}_j)\odot\mathbf{V}_j\right)\left((\mathbf{A}_0\mathbf{W}_{xy,j})\odot\mathbf{V}_j\right)^{\mathrm{T}}\Big)}{t}\\
&\quad - \lim_{t\to 0} \frac{\operatorname{tr}\Big(\left((\mathbf{A}_0\mathbf{W}_{xy,j})\odot\mathbf{V}_j\right)\left((\mathbf{A}_0\mathbf{W}_{xy,j})\odot\mathbf{V}_j\right)^{\mathrm{T}}\Big)}{t}\\
&= -2\lim_{t\to 0} \frac{\operatorname{tr}\Big(t\left((\mathbf{N}_j\mathbf{Y}_j)\odot\mathbf{V}_j\right)\left((\mathbf{K}\mathbf{W}_{xy,j})\odot\mathbf{V}_j\right)^{\mathrm{T}}\Big)}{t}\\
&\quad + 2\lim_{t\to 0} \frac{\operatorname{tr}\Big(t\left((\mathbf{A}_0\mathbf{W}_{xy,j})\odot\mathbf{V}_j\right)\left((\mathbf{K}\mathbf{W}_{xy,j})\odot\mathbf{V}_j\right)^{\mathrm{T}}\Big)}{t}\\
&\quad + \lim_{t\to 0} \frac{\operatorname{tr}\Big(t^2\left((\mathbf{K}\mathbf{W}_{xy,j})\odot\mathbf{V}_j\right)\left((\mathbf{K}\mathbf{W}_{xy,j})\odot\mathbf{V}_j\right)^{\mathrm{T}}\Big)}{t}\\
&= -2\operatorname{tr}\Big(\left((\mathbf{N}_j\mathbf{Y}_j)\odot\mathbf{V}_j\right)\left((\mathbf{K}\mathbf{W}_{xy,j})\odot\mathbf{V}_j\right)^{\mathrm{T}}\Big)\\
&\quad + 2\operatorname{tr}\Big(\left((\mathbf{A}_0\mathbf{W}_{xy,j})\odot\mathbf{V}_j\right)\left((\mathbf{K}\mathbf{W}_{xy,j})\odot\mathbf{V}_j\right)^{\mathrm{T}}\Big)\\
&= 2\operatorname{tr}\Big(\left((\mathbf{A}_0\mathbf{W}_{xy,j} - \mathbf{N}_j\mathbf{Y}_j)\odot\mathbf{V}_j\right)\left((\mathbf{K}\mathbf{W}_{xy,j})\odot\mathbf{V}_j\right)^{\mathrm{T}}\Big)\\
&= 2\operatorname{tr}\Big(\left((\mathbf{K}\mathbf{W}_{xy,j})\odot\mathbf{V}_j\right)\left((\mathbf{A}_0\mathbf{W}_{xy,j} - \mathbf{N}_j\mathbf{Y}_j)\odot\mathbf{V}_j\right)^{\mathrm{T}}\Big)
\end{aligned}$$

$$= 2\,\mathrm{tr}\Big(\left(\mathbf{V}_j \odot (\mathbf{K}\mathbf{W}_{xy,j})\right)\left((\mathbf{A}_0\mathbf{W}_{xy,j} - \mathbf{N}_j\mathbf{Y}_j) \odot \mathbf{V}_j\right)^{\mathrm{T}}\Big)$$
$$= 2\,\mathrm{tr}\Big(\left(\mathbf{V}_j \odot ((\mathbf{A}_0\mathbf{W}_{xy,j} - \mathbf{N}_j\mathbf{Y}_j) \odot \mathbf{V}_j)\right)\left(\mathbf{K}\mathbf{W}_{xy,j}\right)^{\mathrm{T}}\Big)$$
$$= 2\,\mathrm{tr}\Big(\Big((\mathbf{V}_j).^2 \odot (\mathbf{A}_0\mathbf{W}_{xy,j} - \mathbf{N}_j\mathbf{Y}_j)\Big)\mathbf{W}_{xy,j}^{\mathrm{T}}\mathbf{K}^{\mathrm{T}}\Big)$$
$$= 2\,\mathrm{tr}\Big(\mathbf{K}^{\mathrm{T}}\Big((\mathbf{V}_j).^2 \odot (\mathbf{A}_0\mathbf{W}_{xy,j} - \mathbf{N}_j\mathbf{Y}_j)\Big)\mathbf{W}_{xy,j}^{\mathrm{T}}\Big)\,.$$

The gradient is therefore given by

$$\frac{\partial Q}{\partial \mathbf{A}} = 2\left((\mathbf{A}\mathbf{W}_{xy,j} - \mathbf{N}_j\mathbf{Y}_j) \odot (\mathbf{V}_j).^2\right)\mathbf{W}_{xy,j}^{\mathrm{T}}\,,$$

where the operation $(\cdot).^2$ denotes elementwise squaring.

Analogously, the gradient of a summand of (4.45) with respect to $\mathbf{N}_j$ is given by:

$$\begin{aligned}
\frac{\partial Q}{\partial \mathbf{N}_j} &= \frac{\partial}{\partial \mathbf{N}_j}\mathrm{tr}\Big(\left((\mathbf{N}_j\mathbf{Y}_j) \odot \mathbf{V}_j\right)\left((\mathbf{N}_j\mathbf{Y}_j) \odot \mathbf{V}_j\right)^{\mathrm{T}}\Big) \\
&\quad - 2\frac{\partial}{\partial \mathbf{N}_j}\mathrm{tr}\Big(\left((\mathbf{N}_j\mathbf{Y}_j) \odot \mathbf{V}_j\right)\left((\mathbf{A}\mathbf{W}_{xy,j}) \odot \mathbf{V}_j\right)^{\mathrm{T}}\Big) \\
&\quad + \frac{\partial}{\partial \mathbf{N}_j}\mathrm{tr}\Big(\left((\mathbf{A}\mathbf{W}_{xy,j}) \odot \mathbf{V}_j\right)\left((\mathbf{A}\mathbf{W}_{xy,j}) \odot \mathbf{V}_j\right)^{\mathrm{T}}\Big) \\
&= \frac{\partial}{\partial \mathbf{N}_j}\mathrm{tr}\Big(\left((\mathbf{N}_j\mathbf{Y}_j) \odot \mathbf{V}_j\right)\left((\mathbf{N}_j\mathbf{Y}_j) \odot \mathbf{V}_j\right)^{\mathrm{T}}\Big) \\
&\quad - 2\frac{\partial}{\partial \mathbf{N}_j}\mathrm{tr}\Big(\left((\mathbf{N}_j\mathbf{Y}_j) \odot \mathbf{V}_j\right)\left((\mathbf{A}\mathbf{W}_{xy,j}) \odot \mathbf{V}_j\right)^{\mathrm{T}}\Big)
\end{aligned} \tag{A.10}$$

resulting in

$$\begin{aligned}
&D_{\mathbf{K}}f(\mathbf{N}_{j,0}) \\
&= \lim_{t\to 0}\frac{\mathrm{tr}\left(\left(((\mathbf{N}_{j,0} + t\mathbf{K})\mathbf{Y}_j) \odot \mathbf{V}_j\right)\left(((\mathbf{N}_{j,0} + t\mathbf{K})\mathbf{Y}_j) \odot \mathbf{V}_j\right)\right)^{\mathrm{T}}}{t} \\
&\quad - 2\lim_{t\to 0}\frac{\mathrm{tr}\Big(\left(((\mathbf{N}_{j,0} + t\mathbf{K})\mathbf{Y}_j) \odot \mathbf{V}_j\right)\left((\mathbf{A}\mathbf{W}_{xy,j}) \odot \mathbf{V}_j\right)^{\mathrm{T}}\Big)}{t} \\
&\quad - \lim_{t\to 0}\frac{\mathrm{tr}\Big(\left((\mathbf{N}_{j,0}\mathbf{Y}_j) \odot \mathbf{V}_j\right)\left((\mathbf{N}_{j,0}\mathbf{Y}_j) \odot \mathbf{V}_j\right)^{\mathrm{T}}\Big)}{t}
\end{aligned} \tag{A.11}$$

$$
\begin{aligned}
&+ 2 \lim_{t\to 0} \frac{\operatorname{tr}\Big(\left((\mathbf{N}_{j,0}\mathbf{Y}_j) \odot \mathbf{V}_j\right) \left((\mathbf{A}\mathbf{W}_{xy,j}) \odot \mathbf{V}_j\right)^{\mathrm{T}}\Big)}{t} \\
&= 2 \lim_{t\to 0} \frac{\operatorname{tr}\Big(t\left((\mathbf{N}_{j,0}\mathbf{Y}_j) \odot \mathbf{V}_j\right) \left((\mathbf{K}\mathbf{Y}_j) \odot \mathbf{V}_j\right)^{\mathrm{T}}\Big)}{t} \\
&+ \lim_{t\to 0} \frac{\operatorname{tr}\Big(t^2\left((\mathbf{K}\mathbf{Y}_j) \odot \mathbf{V}_j\right) \left((\mathbf{K}\mathbf{Y}_j) \odot \mathbf{V}_j\right)^{\mathrm{T}}\Big)}{t} \\
&- 2 \lim_{t\to 0} \frac{\operatorname{tr}\Big(t\left((\mathbf{K}\mathbf{Y}_j) \odot \mathbf{V}_j\right) \left((\mathbf{A}\mathbf{W}_{xy,j}) \odot \mathbf{V}_j\right)^{\mathrm{T}}\Big)}{t} \\
&= 2 \operatorname{tr}\Big(\left((\mathbf{N}_{j,0}\mathbf{Y}_j) \odot \mathbf{V}_j\right) \left((\mathbf{K}\mathbf{Y}_j) \odot \mathbf{V}_j\right)^{\mathrm{T}}\Big) \\
&- 2 \operatorname{tr}\Big(\left((\mathbf{K}\mathbf{Y}_j) \odot \mathbf{V}_j\right) \left((\mathbf{A}\mathbf{W}_{xy,j}) \odot \mathbf{V}_j\right)^{\mathrm{T}}\Big) \\
&= 2 \operatorname{tr}\Big(\left((\mathbf{N}_{j,0}\mathbf{Y}_j) \odot \mathbf{V}_j\right) \left((\mathbf{K}\mathbf{Y}_j) \odot \mathbf{V}_j\right)^{\mathrm{T}}\Big) \\
&- 2 \operatorname{tr}\Big(\left((\mathbf{A}\mathbf{W}_{xy,j}) \odot \mathbf{V}_j\right) \left((\mathbf{K}\mathbf{Y}_j) \odot \mathbf{V}_j\right)^{\mathrm{T}}\Big) \\
&= 2 \operatorname{tr}\Big(\left((\mathbf{N}_{j,0}\mathbf{Y}_j - \mathbf{A}\mathbf{W}_{xy,j}) \odot \mathbf{V}_j\right) \left((\mathbf{K}\mathbf{Y}_j) \odot \mathbf{V}_j\right)^{\mathrm{T}}\Big) \\
&= 2 \operatorname{tr}\Big(\left((\mathbf{K}\mathbf{Y}_j) \odot \mathbf{V}_j\right) \left((\mathbf{N}_{j,0}\mathbf{Y}_j - \mathbf{A}\mathbf{W}_{xy,j}) \odot \mathbf{V}_j\right)^{\mathrm{T}}\Big) \\
&= 2 \operatorname{tr}\Big(\left(\mathbf{V}_j \odot (\mathbf{K}\mathbf{Y}_j)\right) \left((\mathbf{N}_{j,0}\mathbf{Y}_j - \mathbf{A}\mathbf{W}_{xy,j}) \odot \mathbf{V}_j\right)^{\mathrm{T}}\Big) \\
&= 2 \operatorname{tr}\Big(\left(\mathbf{V}_j \odot ((\mathbf{N}_{j,0}\mathbf{Y}_j - \mathbf{A}\mathbf{W}_{xy,j}) \odot \mathbf{V}_j)\right) (\mathbf{K}\mathbf{Y}_j)^{\mathrm{T}}\Big) \\
&= 2 \operatorname{tr}\Big(\left((\mathbf{V}_j).^2 \odot (\mathbf{N}_{j,0}\mathbf{Y}_j - \mathbf{A}\mathbf{W}_{xy,j})\right) \mathbf{Y}_j^{\mathrm{T}}\mathbf{K}^{\mathrm{T}}\Big) \\
&= 2 \operatorname{tr}\Big(\mathbf{K}^{\mathrm{T}} \left((\mathbf{V}_j).^2 \odot (\mathbf{N}_{j,0}\mathbf{Y}_j - \mathbf{A}\mathbf{W}_{xy,j})\right) \mathbf{Y}_j^{\mathrm{T}}\Big) .
\end{aligned}
$$

The gradient with respect to $\mathbf{N}_j$ is therefore given by

$$
\frac{\partial Q}{\partial \mathbf{N}_j} = 2 \left((\mathbf{N}_j\mathbf{Y}_j - \mathbf{A}\mathbf{W}_{xy,j}) \odot (\mathbf{V}_j).^2 \right) \mathbf{Y}_j^{\mathrm{T}} .
$$

Bibliography

[1] **Ahmed, N., Natarajan, T., and Rao, K. R.** *Discrete cosine transform*. In: *IEEE Transactions on Computers* 100.1 (1974), pp. 90–93.

[2] **Almeida, M. S. and Almeida, L. B.** *Blind and semi-blind deblurring of natural images*. In: *IEEE Transactions on Image Processing* 19.1 (2010), pp. 36–52.

[3] **Amigo, J. M., Babamoradi, H., and Elcoroaristizabal, S.** *Hyperspectral image analysis. A tutorial*. In: *Analytica Chimica Acta* 896.Supplement C (2015), pp. 34–51.

[4] **AVIRIS web page**. URL: https://aviris.jpl.nasa.gov/ (visited on 09/11/2017).

[5] **Ball, G. H. and Hall, D. J.** *ISODATA, a novel method of data analysis and pattern classification*. Tech. rep. Stanford Research Institute, Menlo Park, CA, 1965.

[6] **Begelman, G., Zibulevsky, M., Rivlin, E., and Kolatt, T.** *Blind decomposition of transmission light microscopic hyperspectral cube using sparse representation*. In: *IEEE Transactions on Medical Imaging* 28.8 (2009), pp. 1317–1324.

[7] **Ben-Dor, E., Schläpfer, D., Plaza, A. J., and Malthus, T.** *Hyperspectral remote sensing*. In: *Airborne Measurements for Environmental Research: Methods and Instruments* (2013), pp. 413–456.

[8] **Bergounioux, M. and Piffet, L.** *A second-order model for image denoising*. In: *Set-Valued and Variational Analysis* 18.3-4 (2010), pp. 277–306.

[9] **Bernard, K., Tarabalka, Y., Angulo, J., Chanussot, J., and Benediktsson, J. A.** *Spectral–spatial classification of hyperspectral data based on a stochastic minimum spanning forest approach*. In: *IEEE Transactions on Image Processing* 21.4 (2012), pp. 2008–2021.

[10] **Bertsekas, D. P.** *On the Goldstein-Levitin-Polyak gradient projection method*. In: *IEEE Transactions on Automatic Control* 21.2 (1976), pp. 174–184.

[11] **Bertsekas, D. P.** *Nonlinear programming*. Athena Scientific, 1999.

[12] **Beyerer, J., Puente León, F., and Frese, C.** *Machine Vision – Automated Visual Inspection: Theory, Practice and Applications.* Springer Berlin Heidelberg, 2016.

[13] **Bhattacharya, P. and Burman, P.** *Theory and Methods of Statistics.* Elsevier Science, 2016.

[14] **Bioucas-Dias, J. M.** *A variable splitting augmented Lagrangian approach to linear spectral unmixing.* In: *First Workshop on Hyperspectral Image and Signal Processing: Evolution in Remote Sensing (WHISPERS), 2009*, pp. 1–4.

[15] **Bioucas-Dias, J. M. and Figueiredo, M. A.** *Alternating direction algorithms for constrained sparse regression: Application to hyperspectral unmixing.* In: *2nd Workshop on Hyperspectral Image and Signal Processing: Evolution in Remote Sensing (WHISPERS), 2010.* IEEE. 2010, pp. 1–4.

[16] **Bioucas-Dias, J. M., Plaza, A., Dobigeon, N., Parente, M., Du, Q., Gader, P., and Chanussot, J.** *Hyperspectral Unmixing Overview: Geometrical, Statistical, and Sparse Regression-Based Approaches.* In: *IEEE Journal of Selected Topics in Applied Earth Observations and Remote Sensing* 5.2 (2012), pp. 354–379.

[17] **Bioucas-Dias, J. M., Plaza, A., Camps-Valls, G., Scheunders, P., Nasrabadi, N., and Chanussot, J.** *Hyperspectral remote sensing data analysis and future challenges.* In: *IEEE Geoscience and Remote Sensing Magazine* 1.2 (2013), pp. 6–36.

[18] **Borges, J. S., Bioucas-Dias, J. M., and Marcal, A. R.** *Bayesian hyperspectral image segmentation with discriminative class learning.* In: *IEEE Transactions on Geoscience and Remote Sensing* 49.6 (2011), pp. 2151–2164.

[19] **Boyd, S. P. and Vandenberghe, L.** *Convex optimization.* Cambridge university press, 2004.

[20] **Bredies, K., Kunisch, K., and Pock, T.** *Total generalized variation.* In: *SIAM Journal on Imaging Sciences* 3.3 (2010), pp. 492–526.

[21] **Buades, A., Coll, B., and Morel, J.-M.** *A review of image denoising algorithms, with a new one.* In: *Multiscale Modeling & Simulation* 4.2 (2005), pp. 490–530.

[22] **Camps-Valls, G. and Bruzzone, L.** *Kernel-based methods for hyperspectral image classification.* In: *IEEE Transactions on Geoscience and Remote Sensing* 43.6 (2005), pp. 1351–1362.

[23] **Camps-Valls, G., Tuia, D., Bruzzone, L., and Benediktsson, J. A.** *Advances in hyperspectral image classification: Earth monitoring with statistical learning methods.* In: *IEEE Signal Processing Magazine* 31.1 (2014), pp. 45–54.

[24] **Chan, T.-H., Chi, C.-Y., Huang, Y.-M., and Ma, W.-K.** *A convex analysis-based minimum-volume enclosing simplex algorithm for hyperspectral unmixing*. In: *IEEE Transactions on Signal Processing* 57.11 (2009), pp. 4418–4432.

[25] **Chang, C.** *Hyperspectral Data Processing: Algorithm Design and Analysis*. Wiley, 2013.

[26] **Chang, C.** *Hyperspectral Imaging: Techniques for Spectral Detection and Classification*. Springer US, 2013.

[27] **Charles, A. S., Olshausen, B. A., and Rozell, C. J.** *Learning sparse codes for hyperspectral imagery*. In: *IEEE Journal of Selected Topics in Signal Processing* 5.5 (2011), pp. 963–978.

[28] **Chen, G. and Qian, S.-E.** *Denoising of hyperspectral imagery using principal component analysis and wavelet shrinkage*. In: *IEEE Transactions on Geoscience and Remote Sensing* 49.3 (2011), pp. 973–980.

[29] **Chen, S. S., Donoho, D. L., and Saunders, M. A.** *Atomic decomposition by basis pursuit*. In: *SIAM review* 43.1 (2001), pp. 129–159.

[30] **Chen, Y., Nasrabadi, N. M., and Tran, T. D.** *Hyperspectral image classification using dictionary-based sparse representation*. In: *IEEE Transactions on Geoscience and Remote Sensing* 49.10 (2011), pp. 3973–3985.

[31] **Cichocki, A., Zdunek, R., Phan, A., and Amari, S.** *Nonnegative Matrix and Tensor Factorizations: Applications to Exploratory Multi-way Data Analysis and Blind Source Separation*. Wiley, 2009.

[32] **Consumer Physics**. URL: https://www.consumerphysics.com/ (visited on 12/13/2017).

[33] **Craig, M. D.** *Minimum-volume transforms for remotely sensed data*. In: *IEEE Transactions on Geoscience and Remote Sensing* 32.3 (1994), pp. 542–552.

[34] **Dabov, K., Foi, A., Katkovnik, V., and Egiazarian, K.** *Image denoising by sparse 3-D transform-domain collaborative filtering*. In: *IEEE Transactions on Image Processing* 16.8 (2007), pp. 2080–2095.

[35] **Davenport, M. A., Duarte, M. F., Eldar, Y. C., and Kutyniok, G.** *Introduction to Compressed Sensing*. In: *Compressed Sensing: Theory and Applications*. Ed. by **Eldar, Y. C. and Kutyniok, G.** Oxford: Cambridge University Press, 2012. Chap. 1, pp. 1–68.

[36] **Dennison, P. E., Halligan, K. Q., and Roberts, D. A.** *A comparison of error metrics and constraints for multiple endmember spectral mixture analysis and spectral angle mapper*. In: *Remote Sensing of Environment* 93.3 (2004), pp. 359–367.

[37] **Deutsches Institut für Normung**. *Strahlungsphysik im optischen Bereich und Lichttechnik; Benennung der Wellenlängenbereiche*. DIN5031-7, 1984.

[38] **Dobigeon, N., Tourneret, J.-Y., Richard, C., Bermudez, J. C. M., McLaughlin, S., and Hero, A. O.** *Nonlinear unmixing of hyperspectral images: Models and algorithms*. In: *IEEE Signal Processing Magazine* 31.1 (2014), pp. 82–94.

[39] **Duda, R. O., Hart, P. E., and Stork, D. G.** *Pattern classification*. John Wiley & Sons, 2012.

[40] **Eches, O., Dobigeon, N., and Tourneret, J.-Y.** *Enhancing Hyperspectral Image Unmixing With Spatial Correlations*. In: *IEEE Transactions on Geoscience and Remote Sensing* 49.11 (2011), pp. 4239–4247.

[41] **Efficient Graph-Based Image Segmentation**. URL: http://sourceforge.net/p/gerbil/svn/65/tree/branches/next/seg_felzenszwalb/ (visited on 09/11/2017).

[42] **Elhamifar, E., Sapiro, G., and Vidal, R.** *See all by looking at a few: Sparse modeling for finding representative objects*. In: *IEEE Conference on Computer Vision and Pattern Recognition (CVPR)*. 2012, pp. 1600–1607.

[43] **ElMasry, G., Kamruzzaman, M., Sun, D.-W., and Allen, P.** *Principles and applications of hyperspectral imaging in quality evaluation of agro-food products: a review*. In: *Critical Reviews in Food Science and Nutrition* 52.11 (2012), pp. 999–1023.

[44] **EnMAP web page**. URL: http://www.enmap.org/?q=sensor/ (visited on 09/11/2017).

[45] **European Machine Vision Association, EMVA 1288 standard, Release 3.1**. URL: http://www.emva.org/wp-content/uploads/EMVA1288-3.1a.pdf (visited on 12/07/2017).

[46] **Fauvel, M., Tarabalka, Y., Benediktsson, J. A., Chanussot, J., and Tilton, J. C.** *Advances in spectral-spatial classification of hyperspectral images*. In: *Proceedings of the IEEE* 101.3 (2013), pp. 652–675.

[47] **Felzenszwalb, P. F. and Huttenlocher, D. P.** *Efficient graph-based image segmentation*. In: *International Journal of Computer Vision* 59.2 (2004), pp. 167–181.

[48] **Gillis, N. and Vavasis, S.** *Fast and robust recursive algorithms for separable nonnegative matrix factorization*. In: *IEEE Transactions on Pattern Analysis and Machine Intelligence* 36.4 (2014), pp. 698–714.

[49] **Goetz, A. F.** *Three decades of hyperspectral remote sensing of the Earth: A personal view*. In: *Remote Sensing of Environment* 113 (2009), S5–S16.

[50] **Goetz, A. F., Vane, G., Solomon, J. E., and Rock, B. N.** *Imaging spectrometry for earth remote sensing.* In: *Science* 228.4704 (1985), pp. 1147–1153.

[51] **Golbabaee, M., Arberet, S., and Vandergheynst, P.** *Compressive source separation: theory and methods for hyperspectral imaging.* In: *IEEE Transactions on Image Processing* 22.12 (2013), pp. 5096–5110.

[52] **Green, A. A., Berman, M., Switzer, P., and Craig, M. D.** *A transformation for ordering multispectral data in terms of image quality with implications for noise removal.* In: *IEEE Transactions on Geoscience and Remote Sensing* 26.1 (1988), pp. 65–74.

[53] **Guan, N., Tao, D., Luo, Z., and Yuan, B.** *NeNMF: An optimal gradient method for nonnegative matrix factorization.* In: *IEEE Transactions on Signal Processing* 60.6 (2012), pp. 2882–2898.

[54] **Guo, B., Gunn, S. R., Damper, R. I., and Nelson, J. D.** *Customizing kernel functions for SVM-based hyperspectral image classification.* In: *IEEE Transactions on Image Processing* 17.4 (2008), pp. 622–629.

[55] **Hapke, B.** *Bidirectional reflectance spectroscopy: 1. Theory.* In: *Journal of Geophysical Research: Solid Earth* 86.B4 (1981), pp. 3039–3054.

[56] **Harsanyi, J. C. and Chang, C.-I.** *Hyperspectral image classification and dimensionality reduction: An orthogonal subspace projection approach.* In: *IEEE Transactions on Geoscience and Remote Sensing* 32.4 (1994), pp. 779–785.

[57] **Harville, D.** *Matrix Algebra: Exercises and Solutions.* Springer New York, 2011.

[58] **He, W., Zhang, H., Zhang, L., and Shen, H.** *Hyperspectral image denoising via noise-adjusted iterative low-rank matrix approximation.* In: *IEEE Journal of Selected Topics in Applied Earth Observations and Remote Sensing* 8.6 (2015), pp. 3050–3061.

[59] **Heylen, R., Parente, M., and Gader, P.** *A review of nonlinear hyperspectral unmixing methods.* In: *IEEE Journal of Selected Topics in Applied Earth Observations and Remote Sensing* 7.6 (2014), pp. 1844–1868.

[60] **Hirsch, M., Wareham, R. J., Martin-Fernandez, M. L., Hobson, M. P., and Rolfe, D. J.** *A stochastic model for electron multiplication charge-coupled devices–from theory to practice.* In: *PloS one* 8.1 (2013).

[61] **Hoyer, P. O.** *Non-negative matrix factorization with sparseness constraints.* In: *Journal of machine learning research* 5 (2004), pp. 1457–1469.

[62] **Hu, Y. and Jacob, M.** *Higher degree total variation (HDTV) regularization for image recovery.* In: *IEEE Transactions on Image Processing* 21.5 (2012), pp. 2559–2571.

[63] **Huck, A., Guillaume, M., and Blanc-Talon, J.** *Minimum Dispersion Constrained Nonnegative Matrix Factorization to Unmix Hyperspectral Data*. In: *IEEE Transactions on Geoscience and Remote Sensing* 48.6 (2010), pp. 2590–2602.

[64] **Hyperspectral datasets**. URL: http://www.escience.cn/people/feiyunZHU/Dataset_GT.html (visited on 09/11/2017).

[65] **Hyvärinen, A., Karhunen, J., and Oja, E.** *Independent Component Analysis*. Adaptive and Cognitive Dynamic Systems: Signal Processing, Learning, Communications and Control. Wiley, 2004.

[66] **Image and video denoising by sparse 3D transform-domain collaborative filtering web page**. URL: http://www.cs.tut.fi/~foi/GCF-BM3D/ (visited on 09/11/2017).

[67] **Iordache, M.-D., Bioucas-Dias, J. M., and Plaza, A.** *Sparse unmixing of hyperspectral data*. In: *IEEE Transactions on Geoscience and Remote Sensing* 49.6 (2011), pp. 2014–2039.

[68] **Iordache, M.-D., Bioucas-Dias, J. M., and Plaza, A.** *Total variation spatial regularization for sparse hyperspectral unmixing*. In: *IEEE Transactions on Geoscience and Remote Sensing* 50.11 (2012), pp. 4484–4502.

[69] **Irgenfried, S. and Negara, C.** *A framework for storage, visualization and analysis of multispectral data*. In: *OCM 2013 – Optical Characterization of Materials*. Ed. by **Beyerer, J., Puente León, F., and Längle, T.** Karlsruhe: KIT Scientific Publishing, 2013, pp. 203–214.

[70] **ISODATA implementation**. URL: https://de.mathworks.com/matlabcentral/fileexchange/23350-isodata-segmentation-a-modification-to--jose-suau-rodriguez--code--isodata-m-?focused=5110786&tab=function (visited on 09/11/2017).

[71] **Jähne, B.** *Digital Image Processing*. Springer Berlin Heidelberg, 2005.

[72] **Kelley, C. T.** *Iterative methods for optimization*. SIAM, 1999.

[73] **Keshava, N. and Mustard, J. F.** *Spectral unmixing*. In: *IEEE Signal Processing Magazine* 19.1 (2002), pp. 44–57.

[74] **Kokaly, R. F., Clark, R. N., Swayze, G. A., Livo, K. E., Hoefen, T. M., Pearson, N. C., Wise, R. A., Benzel, W. M., Lowers, H. A., Driscoll, R. L., and Klein, A. J.** *USGS Spectral Library Version 7: U.S. Geological Survey Data Series 1035*. Report. United States Geological Survey, 2017.

[75] **Kruse, F. A., Lefkoff, A., Boardman, J., Heidebrecht, K., Shapiro, A., Barloon, P., and Goetz, A.** *The spectral image processing system (SIPS)—interactive visualization and analysis of imaging spectrometer data*. In: *Remote Sensing of Environment* 44.2-3 (1993), pp. 145–163.

[76] **Lam, A., Sato, I., and Sato, Y.** *Denoising hyperspectral images using spectral domain statistics.* In: *21st International Conference on Pattern Recognition (ICPR).* 2012, pp. 477–480.

[77] **Lee, D. D. and Seung, H. S.** *Algorithms for non-negative matrix factorization.* In: *Advances in Neural Information Processing Systems.* 2001, pp. 556–562.

[78] **Letexier, D. and Bourennane, S.** *Noise removal from hyperspectral images by multidimensional filtering.* In: *IEEE Transactions on Geoscience and Remote Sensing* 46.7 (2008), pp. 2061–2069.

[79] **Li, C., Sun, T., Kelly, K. F., and Zhang, Y.** *A compressive sensing and unmixing scheme for hyperspectral data processing.* In: *IEEE Transactions on Image Processing* 21.3 (2012), pp. 1200–1210.

[80] **Li, J., Bioucas-Dias, J. M., and Plaza, A.** *Spectral–spatial hyperspectral image segmentation using subspace multinomial logistic regression and Markov random fields.* In: *IEEE Transactions on Geoscience and Remote Sensing* 50.3 (2012), pp. 809–823.

[81] **Li, S., Qiu, J., Yang, X., Liu, H., Wan, D., and Zhu, Y.** *A novel approach to hyperspectral band selection based on spectral shape similarity analysis and fast branch and bound search.* In: *Engineering Applications of Artificial Intelligence* 27 (2014), pp. 241–250.

[82] **Liao, W., Goossens, B., Aelterman, J., Luong, H., Pizurica, A., Wouters, N., Saeys, W., and Philips, W.** *Hyperspectral image deblurring with PCA and total variation.* In: *5th IEEE Workshop on Hyperspectral Image and Signal Processing: Evolution in Remote Sensing (WHISPERS).* 2013, pp. 1–4.

[83] **Lin, C.-J.** *Projected gradient methods for nonnegative matrix factorization.* In: *Neural Computation* 19.10 (2007), pp. 2756–2779.

[84] **Lin, T. and Bourennane, S.** *Survey of hyperspectral image denoising methods based on tensor decompositions.* In: *EURASIP Journal on Advances in Signal Processing* 2013.1 (2013), p. 186.

[85] **Loncan, L., Almeida, L. B. de, Bioucas-Dias, J. M., Briottet, X., Chanussot, J., Dobigeon, N., Fabre, S., Liao, W., Licciardi, G. A., Simoes, M., et al.** *Hyperspectral pansharpening: A review.* In: *IEEE Geoscience and Remote Sensing Magazine* 3.3 (2015), pp. 27–46.

[86] **Lu, G. and Fei, B.** *Medical hyperspectral imaging: a review.* In: *Journal of Biomedical Optics* 19.1 (2014), pp. 010901-1–010901-23.

[87] **Lu, R. and Chen, Y.-R.** *Hyperspectral imaging for safety inspection of food and agricultural products.* In: *Photonics East.* International Society for Optics and Photonics. 1999, pp. 121–133.

[88] **Lu, X., Wu, H., and Yuan, Y.** *Double constrained NMF for hyperspectral unmixing*. In: *IEEE Transactions on Geoscience and Remote Sensing* 52.5 (2014), pp. 2746–2758.

[89] **Luisier, F., Blu, T., and Unser, M.** *Image denoising in mixed Poisson–Gaussian noise*. In: *IEEE Transactions on Image Processing* 20.3 (2011), pp. 696–708.

[90] **Lysaker, M., Lundervold, A., and Tai, X.-C.** *Noise removal using fourth-order partial differential equation with applications to medical magnetic resonance images in space and time*. In: *IEEE Transactions on Image Processing* 12.12 (2003), pp. 1579–1590.

[91] **Maggioni, M., Katkovnik, V., Egiazarian, K., and Foi, A.** *Nonlocal transform-domain filter for volumetric data denoising and reconstruction*. In: *IEEE Transactions on Image Processing* 22.1 (2013), pp. 119–133.

[92] **Mäkitalo, M. and Foi, A.** *Optimal inversion of the Anscombe transformation in low-count Poisson image denoising*. In: *IEEE Transactions on Image Processing* 20.1 (2011), pp. 99–109.

[93] **Martín, G. and Plaza, A.** *Region-based spatial preprocessing for endmember extraction and spectral unmixing*. In: *IEEE Geoscience and Remote Sensing Letters* 8.4 (2011), pp. 745–749.

[94] **Mei, S. and He, M.** *Minimum endmember-wise distance constrained nonnegative matrix factorization for spectral mixture analysis of hyperspectral images*. In: *IEEE International Geoscience and Remote Sensing Symposium (IGARSS)*. 2011, pp. 1299–1302.

[95] **Miao, L. and Qi, H.** *Endmember Extraction From Highly Mixed Data Using Minimum Volume Constrained Nonnegative Matrix Factorization*. In: *IEEE Transactions on Geoscience and Remote Sensing* 45.3 (2007), pp. 765–777.

[96] **Michelsburg, M. and Puente León, F.** *Spectral and spatial unmixing for material recognition in sorting plants*. In: *Optical Characterization of Materials*. Ed. by **Beyerer, J., Puente León, F., and Längle, T.** Vol. 1. 2013, pp. 179–190.

[97] **Michelsburg, M., Gruna, R., Vieth, K.-U., and Puente León, F.** *Spektrale Bandselektion für das Filterdesign optischer Inspektionssysteme*. In: *tm–Technisches Messen* 78.9 (2011), pp. 384–390.

[98] **Mielikainen, J. and Toivanen, P.** *Clustered DPCM for the lossless compression of hyperspectral images*. In: *IEEE Transactions on Geoscience and Remote Sensing* 41.12 (2003), pp. 2943–2946.

[99] **Mitchell, P. A.** *Hyperspectral digital imagery collection experiment (HYDICE)*. In: *Geographic Information Systems, Photogrammetry, and Geological/Geophysical Remote Sensing*. Vol. 2587. Proceedings of SPIE. 1995.

[100] **Mittag, H.-J. and Rinne, H.** *Statistical Methods of Quality Assurance.* Taylor & Francis, 1993.

[101] **MultiSpec–Hyperspectral images web page.** URL: https://engineering.purdue.edu/~biehl/MultiSpec/hyperspectral.html (visited on 09/11/2017).

[102] **NASA's RHSEG Common Software Package.** URL: http://opensource.gsfc.nasa.gov/projects/HSEG/ (visited on 09/11/2017).

[103] **Nascimento, J. M. P. and Bioucas-Dias, J. M.** *Nonlinear mixture model for hyperspectral unmixing.* In: *Image and Signal Processing for Remote Sensing.* Vol. 7477. Proceedings of SPIE. 2009.

[104] **Nasrabadi, N. M.** *Hyperspectral target detection: An overview of current and future challenges.* In: *IEEE Signal Processing Magazine* 31.1 (2014), pp. 34–44.

[105] **Nocedal, J. and Wright, S. J.** *Numerical optimization.* Vol. 2. Springer New York, 1999.

[106] **Othman, H. and Qian, S.-E.** *Noise reduction of hyperspectral imagery using hybrid spatial-spectral derivative-domain wavelet shrinkage.* In: *IEEE Transactions on Geoscience and Remote Sensing* 44.2 (2006), pp. 397–408.

[107] **Paoli, A., Melgani, F., and Pasolli, E.** *Clustering of hyperspectral images based on multiobjective particle swarm optimization.* In: *IEEE Transactions on Geoscience and Remote Sensing* 47.12 (2009), pp. 4175–4188.

[108] **Papadopoulo, T. and Lourakis, M. I.** *Estimating the Jacobian of the singular value decomposition: Theory and applications.* In: *European Conference on Computer Vision.* Springer. 2000, pp. 554–570.

[109] **Parente, M. and Zymnis, A.** *Statistical clustering and mineral spectral unmixing in AVIRIS hyperspectral image of Cuprite, NV.* In: *CS2009 report* (2005).

[110] **Perona, P. and Malik, J.** *Scale-space and edge detection using anisotropic diffusion.* In: *IEEE Transactions on Pattern Analysis and Machine Intelligence* 12.7 (1990), pp. 629–639.

[111] **Petersen, K. B. and Pedersen, M. S.** *The matrix cookbook.* Technical University of Denmark, 2008.

[112] **Piqueras, S., Duponchel, L., Tauler, R., and Juan, A. de.** *Resolution and segmentation of hyperspectral biomedical images by Multivariate Curve Resolution-Alternating Least Squares.* In: *Analytica Chimica Acta* 705.1–2 (2011), pp. 182–192.

[113] **Plaza, A., Martínez, P., Pérez, R., and Plaza, J.** *Spatial/spectral endmember extraction by multidimensional morphological operations.* In: *IEEE Transactions on Geoscience and Remote Sensing* 40.9 (2002), pp. 2025–2041.

[114] **Plaza, A., Martínez, P., Pérez, R., and Plaza, J.** *A quantitative and comparative analysis of endmember extraction algorithms from hyperspectral data.* In: *IEEE Transactions on Geoscience and Remote Sensing* 42.3 (2004), pp. 650–663.

[115] **Plaza, A., Benediktsson, J. A., Boardman, J. W., Brazile, J., Bruzzone, L., Camps-Valls, G., Chanussot, J., Fauvel, M., Gamba, P., Gualtieri, A., et al.** *Recent advances in techniques for hyperspectral image processing.* In: *Remote Sensing of Environment* 113 (2009), S110–S122.

[116] **Plaza, A., Martín, G., Plaza, J., Zortea, M., and Sánchez, S.** *Recent Developments in Endmember Extraction and Spectral Unmixing.* In: *Optical Remote Sensing: Advances in Signal Processing and Exploitation Techniques.* Ed. by **Prasad, S., Bruce, L. M., and Chanussot, J.** Berlin, Heidelberg: Springer Berlin Heidelberg, 2011, pp. 235–267.

[117] **Polyak, B. T.** *Introduction to Optimization.* Optimization Software, New York, 1987.

[118] **PQN implementation.** URL: www.cs.ubc.ca/~schmidtm/Software/PQN.html (visited on 09/11/2017).

[119] **Qi, L. and Womersley, R. S.** *On extreme singular values of matrix valued functions.* In: *Journal of Convex Analysis* 3 (1996), pp. 153–166.

[120] **Qian, Y., Jia, S., Zhou, J., and Robles-Kelly, A.** *Hyperspectral unmixing via sparsity-constrained nonnegative matrix factorization.* In: *IEEE Transactions on Geoscience and Remote Sensing* 49.11 (2011), pp. 4282–4297.

[121] **Qian, Y., Shen, Y., Ye, M., and Wang, Q.** *3-D nonlocal means filter with noise estimation for hyperspectral imagery denoising.* In: *IEEE International Geoscience and Remote Sensing Symposium (IGARSS).* 2012, pp. 1345–1348.

[122] **Rajabi, R. and Ghassemian, H.** *Spectral unmixing of hyperspectral imagery using multilayer NMF.* In: *IEEE Geoscience and Remote Sensing Letters* 12.1 (2015), pp. 38–42.

[123] **Ravishanker, N. and Dey, D. K.** *A First Course in Linear Model Theory.* Chapman & Hall/CRC, 2001.

[124] **Rodríguez, P. and Wohlberg, B.** *Efficient minimization method for a generalized total variation functional.* In: *IEEE Transactions on Image Processing* 18.2 (2009), pp. 322–332.

[125] **Roggo, Y., Edmond, A., Chalus, P., and Ulmschneider, M.** *Infrared hyperspectral imaging for qualitative analysis of pharmaceutical solid forms.* In: *Analytica Chimica Acta* 535.1 (2005), pp. 79–87.

[126] **Rudas, T.** *Handbook of Probability: Theory and Applications.* SAGE Publications, 2008.

[127] **Rudin, L. I., Osher, S., and Fatemi, E.** *Nonlinear total variation based noise removal algorithms.* In: *Physica D: Nonlinear Phenomena* 60.1–4 (1992), pp. 259–268.

[128] **Rump, S. M.** *Verified bounds for singular values, in particular for the spectral norm of a matrix and its inverse.* In: *BIT Numerical Mathematics* 51.2 (2011), pp. 367–384.

[129] **Schmidt, M. W., Berg, E. van den, Friedlander, M. P., and Murphy, K.** *Optimizing costly functions with simple constraints: A limited-memory projected quasi-Newton algorithm.* In: *International Conference on Artificial Intelligence and Statistics.* 2009, pp. 456–463.

[130] **Schönemann, P. H.** *On the formal differentiation of traces and determinants.* In: *Multivariate Behavioral Research* 20.2 (1985), pp. 113–139.

[131] **Shi, C. and Wang, L.** *Incorporating spatial information in spectral unmixing: A review.* In: *Remote Sensing of Environment* 149 (2014), pp. 70–87.

[132] **Shor, N. Z., Kiwiel, K. C., and Ruszczynski, A.** *Minimization Methods for Non-Differentiable Functions.* Springer Series in Computational Mathematics. Springer Berlin Heidelberg, 2012.

[133] **Somers, B., Asner, G. P., Tits, L., and Coppin, P.** *Endmember variability in spectral mixture analysis: A review.* In: *Remote Sensing of Environment* 115.7 (2011), pp. 1603–1616.

[134] **Song, X., Jiang, X., and Rui, X.** *Spectral unmixing using linear unmixing under spatial autocorrelation constraints.* In: *IEEE International Geoscience and Remote Sensing Symposium (IGARSS).* 2010, pp. 975–978.

[135] **Stedwell, C. N. and Polfer, N. C.** *Spectroscopy and the Electromagnetic Spectrum.* In: *Laser Photodissociation and Spectroscopy of Mass-separated Biomolecular Ions. Lecture Notes in Chemistry.* Vol. 83. Springer, 2013, pp. 1–20.

[136] **Strichartz, R. S.** *A Guide to Distribution Theory and Fourier Transforms.* World Scientific, 2003.

[137] **Sun, D.-W.** *Hyperspectral imaging for food quality analysis and control.* Elsevier, 2010.

[138] **Takayanagi, M.** *Near-Infrared Spectroscopy*. In: *Introduction to Experimental Infrared Spectroscopy: Fundamentals and Practical Methods*. Wiley, 2014, pp. 253–268.

[139] **Tam, T.-Y.** *A. Horn's result on matrices with prescribed singular values and eigenvalues*. In: *Electronic J. Linear Algebra* 21 (2010), pp. 25–27.

[140] **Tarabalka, Y., Benediktsson, J. A., and Chanussot, J.** *Spectral–spatial classification of hyperspectral imagery based on partitional clustering techniques*. In: *IEEE Transactions on Geoscience and Remote Sensing* 47.8 (2009), pp. 2973–2987.

[141] **Themelis, K. E., Rontogiannis, A. A., and Koutroumbas, K. D.** *A novel hierarchical Bayesian approach for sparse semisupervised hyperspectral unmixing*. In: *IEEE Transactions on Signal Processing* 60.2 (2012), pp. 585–599.

[142] **Thenkabail, P. S. and Lyon, J. G.** *Hyperspectral remote sensing of vegetation*. CRC Press, 2016.

[143] **Tilton, J. C.** *Image segmentation by region growing and spectral clustering with natural convergence criterion*. In: *IEEE International Geoscience and Remote Sensing Symposium (IGARSS)*. 1998, pp. 1766–1768.

[144] **Tilton, J. C.** *Analysis of hierarchically related image segmentations*. In: *IEEE Workshop on Advances in Techniques for Analysis of Remotely Sensed Data, 2003*, pp. 60–69.

[145] **USGS Spectral Library**. URL: https://speclab.cr.usgs.gov/spectral-lib.html (visited on 12/18/2017).

[146] **Ushakov, N.** *Selected Topics in Characteristic Functions*. VSP, 1999.

[147] **VTT optical microspectrometer**. URL: http://www.vttresearch.com/services/smart-industry/space-technologies/sensors-imaging-and-data-analysis/optical-microspectrometer (visited on 12/13/2017).

[148] **Wang, Y., Niu, R., and Yu, X.** *Anisotropic diffusion for hyperspectral imagery enhancement*. In: *IEEE Sensors Journal* 10.3 (2010), pp. 469–477.

[149] **Weyl, H.** *Inequalities between the two kinds of eigenvalues of a linear transformation*. In: *Proceedings of the National Academy of Sciences* 35.7 (1949), pp. 408–411.

[150] **Winter, M. E.** *N-FINDR: An algorithm for fast autonomous spectral endmember determination in hyperspectral data*. In: *Imaging Spectrometry*. Vol. 3753. Proceedings of SPIE. 1999, pp. 266–276.

[151] **Xu, L., Li, F., Wong, A., and Clausi, D.** *Hyperspectral image denoising using a spatial–spectral Monte Carlo sampling approach.* In: *IEEE Journal of Selected Topics in Applied Earth Observations and Remote Sensing* 8.6 (2015), pp. 3025–3038.

[152] **Yang, F., Lu, Y. M., Sbaiz, L., and Vetterli, M.** *Bits from photons: Oversampled image acquisition using binary Poisson statistics.* In: *IEEE Transactions on Image Processing* 21.4 (2012), pp. 1421–1436.

[153] **Yang, Z., Zhou, G., Xie, S., Ding, S., Yang, J.-M., and Zhang, J.** *Blind spectral unmixing based on sparse nonnegative matrix factorization.* In: *IEEE Transactions on Image Processing* 20.4 (2011), pp. 1112–1125.

[154] **Yu, Y., Guo, S., and Sun, W.** *Minimum distance constrained non-negative matrix factorization for the endmember extraction of hyperspectral images.* In: *Remote Sensing and GIS Data Processing and Applications; and Innovative Multispectral Technology and Applications.* Vol. 6790. Proceedings of SPIE. 2007.

[155] **Yuan, Q., Zhang, L., and Shen, H.** *Hyperspectral image denoising employing a spectral–spatial adaptive total variation model.* In: *IEEE Transactions on Geoscience and Remote Sensing* 50.10 (2012), pp. 3660–3677.

[156] **Zare, A., Gader, P., and Gurumoorthy, K. S.** *Directly measuring material proportions using hyperspectral compressive sensing.* In: *IEEE Geoscience and Remote Sensing Letters* 9.3 (2012), pp. 323–327.

[157] **Zare, A. and Ho, K.** *Endmember variability in hyperspectral analysis: Addressing spectral variability during spectral unmixing.* In: *IEEE Signal Processing Magazine* 31.1 (2014), pp. 95–104.

[158] **Zhang, H.** *Hyperspectral image denoising with cubic total variation model.* In: *ISPRS Annals of Photogrammetry, Remote Sensing and Spatial Information Sciences* 7 (2012), pp. 95–98.

[159] **Zhang, H., He, W., Zhang, L., Shen, H., and Yuan, Q.** *Hyperspectral image restoration using low-rank matrix recovery.* In: *IEEE Transactions on Geoscience and Remote Sensing* 52.8 (2014), pp. 4729–4743.

[160] **Zhu, F., Wang, Y., Xiang, S., Fan, B., and Pan, C.** *Structured Sparse Method for Hyperspectral Unmixing.* In: *ISPRS Journal of Photogrammetry and Remote Sensing* 88 (2014), pp. 101–118.

[161] **Zibulevsky, M. and Pearlmutter, B. A.** *Blind source separation by sparse decomposition in a signal dictionary.* In: *Neural Computation* 13.4 (2001), pp. 863–882.

[162] **Zortea, M. and Plaza, A.** *Spatial Preprocessing for Endmember Extraction.* In: *IEEE Transactions on Geoscience and Remote Sensing* 47.8 (2009), pp. 2679–2693.

[163] **Zymnis, A., Kim, S.-J., Skaf, J., Parente, M., and Boyd, S.** *Hyperspectral Image Unmixing via Alternating Projected Subgradients.* In: *Conference Record of the Forty-First Asilomar Conference on Signals, Systems and Computers, ACSSC.* 2007, pp. 1164–1168.

List of Publications

[164] **Bauer, S., Mackens, P., and Puente León, F.** *Entrauschungsalgorithmus für Hyperspektralbilder mit Poisson-Statistik.* In: *XXIX. Messtechnisches Symposium 2015.* Ed. by **Fröhlich, T. and Manske, E.** Berlin: De Gruyter Oldenbourg, 2015, pp. 91–98.

[165] **Bauer, S., Mann, D., and Puente León, F.** *Applicability of hyperspectral fluorescence imaging for mineral sorting.* In: *OCM 2015 - Optical Characterization of Materials.* Ed. by **Beyerer, J., Puente León, F., and Längle, T.** Karlsruhe: KIT Scientific Publishing, 2015, pp. 205–214.

[166] **Bauer, S., Neumann, F., and Puente León, F.** *Spatial regularization for the unmixing of hyperspectral images.* In: *Automated Visual Inspection and Machine Vision.* Vol. 9530. Proceedings of SPIE. 2015.

[167] **Bauer, S. and Puente León, F.** *Industrielle Sortierung von Mineralen anhand von hyperspektralen Fluoreszenzaufnahmen – Potenzialbewertung.* In: *Forum Bildverarbeitung 2014.* Ed. by **Puente León, F. and Heizmann, M.** KIT Scientific Publishing, 2014, pp. 215–226.

[168] **Bauer, S. and Puente León, F.** *Gewinnung und Verarbeitung hyperspektraler Fluoreszenzbilder zur optischen Mineralklassifikation.* In: *tm – Technisches Messen* 82.1 (2015), pp. 24–33.

[169] **Bauer, S. and Puente León, F.** *Spectral and geometric aspects of mineral identification by means of hyperspectral fluorescence imaging.* In: *tm – Technisches Messen* 82.12 (2015), pp. 597–605.

[170] **Bauer, S. and Puente León, F.** *Hyperspectral fluorescence data fusion using quaternion and octonion phase.* In: *IEEE International Conference on Multisensor Fusion and Integration.* Baden-Baden, 2016, pp. 615–621.

[171] **Bauer, S. and Puente León, F.** *Using image pyramids for the acceleration of spectral unmixing based on nonnegative matrix factorization.* In: *8th IEEE Workshop on Hyperspectral Image and Signal Processing: Evolution in Remote Sensing (WHISPERS).* Los Angeles, CA, 2016, pp. 733–737.

[172] **Bauer, S. and Puente León, F.** *Vierdimensionale Mineralerkennung mittels hyperspektraler Fluoreszenzbilder.* In: *Sensoren und Messsysteme 2016.* Nürnberg, 2016, pp. 78–83.

[173] **Bauer, S., Stefan, J., and Puente León, F.** *Hyperspectral image unmixing involving spatial information by extending the alternating least-squares algorithm.* In: *tm – Technisches Messen* 82.4 (2015), pp. 174–186.

[174] **Bauer, S., Suchaneck, A., and Puente León, F.** *Thermal and energy battery management optimization in electric vehicles using Pontryagin's maximum principle.* In: *Journal of Power Sources* 246 (2014), pp. 808–818.

[175] **Bauer, S., Winterbauer, E., and Puente León, F.** *Accelerating spectral unmixing by using clustered images.* In: *IEEE International Conference on Image Processing (ICIP).* Beijing, China, 2017.

[176] **Bauer, S., Stefan, J., Michelsburg, M., Längle, T., and Puente León, F.** *Robustness improvement of hyperspectral image unmixing by spatial second-order regularization.* In: *IEEE Transactions on Image Processing* 23.12 (2014), pp. 5209–5221.

[177] **Bauer, S., Stefan, J., Michelsburg, M., and Puente León, F.** *Verbesserung der Entmischung hyperspektraler Bilder durch den Einbezug räumlicher Information.* In: *XXVIII. Messtechnisches Symposium des Arbeitskreises der Hochschullehrer für Messtechnik e.V.* Ed. by **Schütze, A.** Shaker Verlag, Aachen, 2014, pp. 7–14.

[178] **Bauer, S., Krippner, W., Luo, D., Taphanel, M., Abdo, M., Badilita, V., Längle, T., Beyerer, J., Korvink, J., and Puente León, F.** *Ortsaufgelöste optische Bestimmung von Materialanteilen in Mischungen.* In: *XXX. Messtechnisches Symposium 2016.* Ed. by **Zimmermann, S.** De Gruyter, Berlin, 2016, pp. 69–76.

[179] **Krippner, W., Bauer, S., and Puente León, F.** *Considering spectral variability for optical material abundance estimation.* In: *tm – Technisches Messen* (2017).

[180] **Krippner, W., Bauer, S., and Puente León, F.** *Optical measurement of material abundances in mixtures incorporating preprocessing to mitigate spectral variability.* In: *OCM 2017 – Optical Characterization of Materials.* KIT Scientific Publishing, Karlsruhe, 2017, pp. 87–96.

[181] **Krippner, W., Bauer, S., and Puente León, F.** *Ortsaufgelöste optische Bestimmung von Materialanteilen in Mischungen*. In: *tm – Technisches Messen* 84.3 (2017), pp. 207–215.

[182] **Krippner, W., Wagner, F., Bauer, S., and Puente León, F.** *Optical determination of material abundances by using neural networks for the derivation of spectral filters*. In: *Automated Visual Inspection and Machine Vision II*. Ed. by **Beyerer, J. and Puente León, F.** Vol. 10334. Proceedings of SPIE. 2017.

[183] **Luo, D., Taphanel, M., Bauer, S., Längle, T., Puente León, F., and Beyerer, J.** *Optical unmixing using programmable spectral source based on DMD*. In: *Next-Generation Spectroscopic Technologies*. Ed. by **Druy, M. A. and Crocombe, R. A.** Vol. 9855. Proceedings of SPIE. 2016.

[184] **Puente León, F. and Bauer, S.** *Praxis der Digitalen Signalverarbeitung*. KIT Scientific Publishing, Karlsruhe, 2015.

[185] **Puente León, F. and Bauer, S.** *Praxis der Digitalen Signalverarbeitung*. 2nd ed. KIT Scientific Publishing, Karlsruhe, 2017.

List of Supervised Theses

[186] **Albayrak, G.** *Nutzung von Salienzenkarten für die vierdimensionale Analyse von hyperspektralen Fluoreszenzbildern*. Diploma thesis. KIT, 2015.

[187] **Ayadi, R.** *Mineralklassifikation mittels hyperspektraler Fluoreszenzbilder im sichtbaren Lichtbereich*. Bachelor thesis. KIT, 2014.

[188] **Bockaj, Z.** *Design of an Interference Filter for Grape-Sorting*. Master thesis. KIT, 2015.

[189] **Bouazizi, A.** *Untersuchung verschiedener Verfahren zum Unmixing hyperspektraler Bilder*. Bachelor thesis. KIT, 2015.

[190] **Eckstein, V.** *Klassifikation von Mineralen durch Auswertung von Bildern ihrer Koronaentladung*. Bachelor thesis. KIT, 2016.

[191] **Erük, M. E.** *Human Activity Recognition for Wearables Using Machine Learning Techniques*. Master thesis. KIT, 2016.

[192] **Forreiter, T.** *Anwendung der Multivariate Curve Resolution auf hyperspektrale Bilder zur optischen Unterscheidung von Substanzen*. Diploma thesis. KIT, 2015.

[193] **Huber, S.** *Mineralklassifikation mittels spektraler Analyse von Koronaentladungen*. Bachelor thesis. KIT, 2015.

[194] **Mackens, P.** *Bildverbesserung, Analyse und Klassifikation von hyperspektralen Fluoreszenzbildern*. Master thesis. KIT, 2015.

[195] **Mann, D.** *Objektklassifikation mittels hyperspektraler Fluoreszenzbilder im sichtbaren Lichtbereich*. Diploma thesis. KIT, 2015.

[196] **Marek, M.** *Sharing Real-World Context between Driver and Vehicle in Spoken Dialogue*. Bachelor thesis. KIT, 2014.

[197] **Neumann, F.** *Untersuchung verschiedener Matrixnormen zum regularisierten Spectral Unmixing hyperspektraler Bilder*. Master thesis. KIT, 2014.

[198] **Phan, M. H.** *Einbezug von Quaternionen in die Entmischung hyperspektraler Bilder*. Master thesis. KIT, 2016.

[199] **Rempel, A.** *Implementierung und Untersuchung von Algorithmen zur Extraktion von Reinspektren aus Hyperspektralbildern*. Bachelor thesis. KIT, 2015.

[200] **Stefan, J.** *Anwendung von Multiway-Datenanalyse auf hyperspektrale Bilder zur optischen Unterscheidung von Substanzen*. Master thesis. KIT, 2013.

[201] **Steker, S.** *Spektrale Entmischung hyperspektraler Bilder mit räumlicher und spektraler Regularisierung*. Master thesis. KIT, 2015.

[202] **Wagner, F.** *Klassifikation von Mineralen anhand von Bildern ihrer Koronaentladung*. Bachelor thesis. KIT, 2014.

[203] **Winterbauer, E.** *Beschleunigte Entmischung hyperspektraler Bilder*. Master thesis. KIT, 2015.

Forschungsberichte aus der Industriellen Informationstechnik (ISSN 2190-6629)

Institut für Industrielle Informationstechnik
Karlsruher Institut für Technologie (KIT)

Hrsg.: Prof. Dr.-Ing. Fernando Puente León, Prof. Dr.-Ing. habil. Klaus Dostert

Band 1 Pérez Grassi, Ana
Variable illumination and invariant features for detecting and classifying varnish defects. (2010)
ISBN 978-3-86644-537-6

Band 2 Christ, Konrad
Kalibrierung von Magnet-Injektoren für Benzin-Direkteinspritzsysteme mittels Körperschall. (2011)
ISBN 978-3-86644-718-9

Band 3 Sandmair, Andreas
Konzepte zur Trennung von Sprachsignalen in unterbestimmten Szenarien. (2011)
ISBN 978-3-86644-744-8

Band 4 Bauer, Michael
Vergleich von Mehrträger-Übertragungsverfahren und Entwurfskriterien für neuartige Powerline-Kommunikationssysteme zur Realisierung von Smart Grids. (2012)
ISBN 978-3-86644-779-0

Band 5 Kruse, Marco
Mehrobjekt-Zustandsschätzung mit verteilten Sensorträgern am Beispiel der Umfeldwahrnehmung im Straßenverkehr. (2013)
ISBN 978-3-86644-982-4

Band 6 Dudeck, Sven
Kamerabasierte In-situ-Überwachung gepulster Laserschweißprozesse. (2013)
ISBN 978-3-7315-0019-3

Band 7 Liu, Wenqing
Emulation of Narrowband Powerline Data Transmission Channels and Evaluation of PLC Systems. (2013)
ISBN 978-3-7315-0071-1

Die Bände sind unter www.ksp.kit.edu als PDF frei verfügbar oder als Druckausgabe bestellbar.

Band 8 Otto, Carola
Fusion of Data from Heterogeneous Sensors with Distributed Fields of View and Situation Evaluation for Advanced Driver Assistance Systems. (2013)
ISBN 978-3-7315-0073-5

Band 9 Wang, Limeng
Image Analysis and Evaluation of Cylinder Bore Surfaces in Micrographs. (2014)
ISBN 978-3-7315-0239-5

Band 10 Michelsburg, Matthias
Materialklassifikation in optischen Inspektionssystemen mithilfe hyperspektraler Daten. (2014)
ISBN 978-3-7315-0273-9

Band 11 Pallauf, Johannes
Objektsensitive Verfolgung und Klassifikation von Fußgängern mit verteilten Multi-Sensor-Trägern. (2016)
ISBN 978-3-7315-0529-7

Band 12 Sigle, Martin
Robuste Schmalband-Powerline-Kommunikation für Niederspannungsverteilernetze. (2016)
ISBN 978-3-7315-0539-6

Band 13 Opalko, Oliver
Powerline-Kommunikation für Batteriemanagement-Systeme in Elektro- und Hybridfahrzeugen. (2017)
ISBN 978-3-7315-0647-8

Band 14 Han, Bin
Characterization and Emulation of Low-Voltage Power Line Channels for Narrowband and Broadband Communication. (2017)
ISBN 978-3-7315-0654-6

Band 15 Alonso, Damián Ezequiel
Wireless Data Transmission for the Battery Management System of Electric and Hybrid Vehicles. (2017)
ISBN 978-3-7315-0670-6

Band 16 Hernández Mesa, Pilar
Design and analysis of a content-based image retrieval system. (2017)
ISBN 978-3-7315-0692-8

Band 17 Suchaneck, André
Energiemanagement-Strategien für batterieelektrische Fahrzeuge. (2018)
ISBN 978-3-7315-0773-4

Band 18 Bauer, Sebastian
Hyperspectral Image Unmixing Incorporating Adjacency Information. (2018)
ISBN 978-3-7315-0788-8